CITY IN THE WOODS

CITY IN THE WOODS

The Life and Design of an American Camp Meeting on Martha's Vineyard

ELLEN WEISS

New York Oxford
OXFORD UNIVERSITY PRESS
1987

Oxford University Press

Oxford New York Toronto
Delhi Bombay Calcutta Madras Karachi
Petaling Jaya Singapore Hong Kong Tokyo
Nairobi Dar es Salaam Cape Town
Melbourne Auckland

and associated companies in
Beirut Berlin Ibadan Nicosia

Published by Oxford University Press, Inc.
200 Madison Avenue, New York, New York 10016

Library of Congress Cataloging-in-Publication Data

Weiss, Ellen (Ellen B.)
City in the woods.

Revision of the author's thesis (Ph. D.)—
University of Illinois at Urbana-Champaign.
Bibliography: p. Includes index.
1. Camp—meetings—Massachusetts—Oak Bluffs.
2. Summer resorts—Massachusetts—Oak Bluffs.
3. Oak Bluffs (Mass.)—History. I. Title
BX8476.016W44 1987 711'.4'0974494 86-8601
ISBN 0-19-504163-1 (alk. paper)

1 3 5 7 9 8 6 4 2

Printed in the United States of America
on acid-free paper

For my parents

ACKNOWLEDGMENTS

It is always a happy task to note, with gratitude, some of the help received during the course of labor. The staff of the Gale Huntington Library of the Dukes County Historical Society took the brunt of the author's demands over many years. Work was also accomplished at the library of the University of Illinois, Urbana-Champaign, the Rhode Island Historical Society Library, the New Bedford Free Library, the Loeb Library at Harvard University, the New England Methodist Historical Society Library at Boston University, and the Boston Public Library.

Among the many individuals who helped are W. T. Tripp, Lydia Swift, Jill Bouck, Christine Stoddard, Marshall Cook, Babs Brittain, Stephanie Michalczyk, Arthur Railton, Phillip and Anita Buddington, Judith Wolin, Julian Weiss, Kathy Reed, and my editor, Joyce Berry. I owe much to the criticism of my dissertation committee at the University of Illinois, Professors John S. Garner, Winton U. Solberg, and especially its chairman, Walter L. Creese. His inspiration and instruction, early and late, have been fundamental. Without his support, along with that of the late Henry Beetle Hough, Edith Blake, and my parents, Harry and Gertrude S. Weiss, this book would not have been written.

CONTENTS

The truth is, human life needs to be dotted over with occasions of stirring interest. The journey asks its milestones, or rather, if you please, its watering places along the way. Our nature requires the recurrence now and then of some event of special interest; something that shall peer up from the dead level of existence—an object for hope to rest upon in the future—an oasis in the desert of the remembered past.

B. W. GORHAM, Camp Meeting Manual, 1854

We bless the Lord for camp meetings and their surroundings—more especially for the surroundings.

New Bedford Mercury, 23 August 1864

INTRODUCTION

Wesleyan Grove is a Methodist camp meeting founded in 1835 on the island of Martha's Vineyard, Massachusetts. A planned resort, Oak Bluffs, was laid out in 1867 next to the by then famous camp meeting. These two extraordinary communities were built to several sets of unique forms in architecture and in plan, making on the land a peculiar place, a magical environment, a work of art. How this happened and how it was observed in its own time is the subject of this book.

Entering the 34-acre precinct of Wesleyan Grove is not very different today than it was in the nineteenth century. People now, as then, are surprised to find themselves in a world which is so unlike any of their previous experience that it seems they have stumbled upon the dwellings of an unknown people, an archaeological mystery whose only evidence is the lost society's strange built remains. Most of the buildings are narrow two-story cottages of an odd proportion and scale which create, when seen in quantity, a hallucinatory sense of otherworldliness. These cottages differ from any others in structural and constructional means as well as in appearance—usually striking observers as absurd comments on the notion of "house" rather than as serious houses themselves. They constitute an independent American building type which was probably invented at this camp meeting by island carpenters. In 1880 there were about 500 of these buildings. About 300 remain today.

Along with and integrated into the peculiar nature of the cottages is their arrangement on the land. They sit cheek by jowl on lots so tiny that there is no room for private outdoor space other than front porches. Wide double doors expose most of the interior and open directly onto public paths or variously shaped small parks, creating a radically different pattern of public and private space from any other single family residential community. The dozen or so cottage-rimmed small parks surround a broad

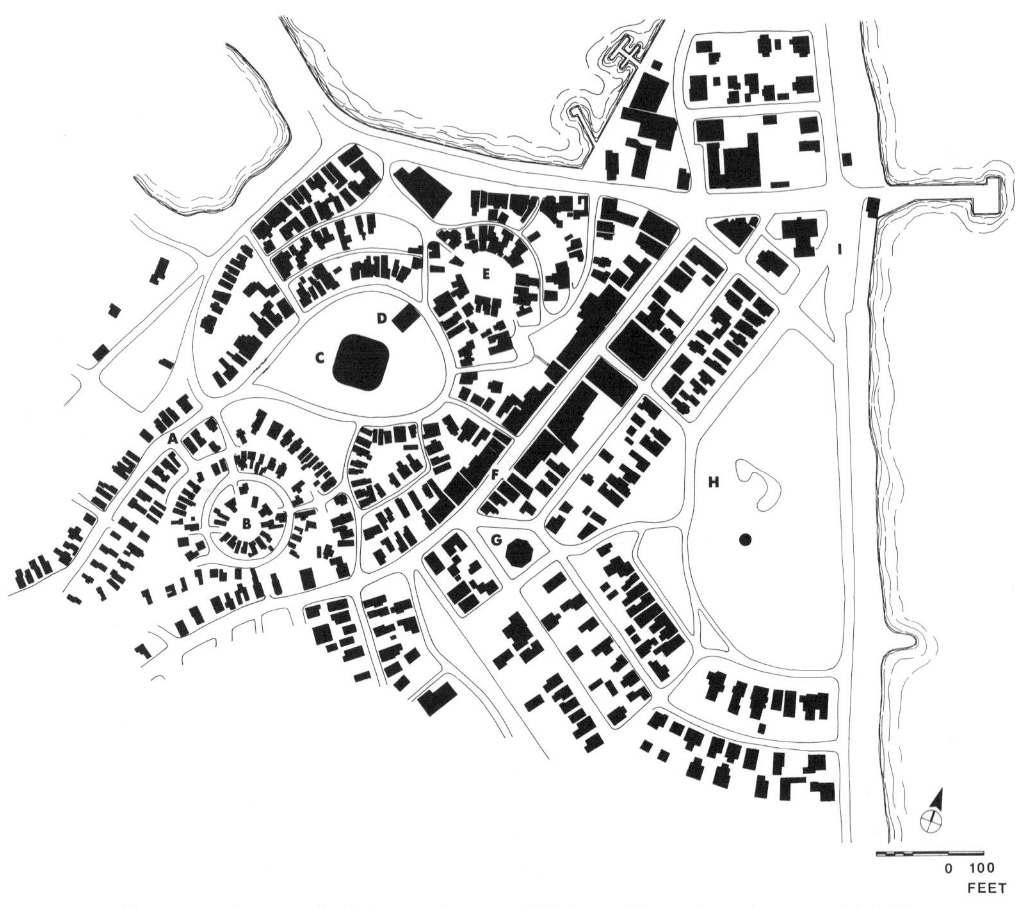

Wesleyan Grove: A. Clinton Avenue; B. Forest Circle; C. tabernacle; D. Trinity Church; E. County Street Park

Oak Bluffs: F. Circuit Avenue; G. Union Chapel; H. Ocean Park; I. Site of Sea View hotel

Wesleyan Grove (left) and Oak Bluffs (right) in modern times, figure-ground plan. (Ronald J. Melvin)

5-acre open area in the precinct's center and this, in turn, contains a vast iron tabernacle, a shed about 140 feet in diameter, which shelters the original consecrated preaching area and which climaxes the huddles of little cottages, like the sun organizes the planets.

By any standards of Western tradition, this is eccentric place-making, clearly implying an unusual relationship between family and community. In

Campground cottage. (Dukes County Historical Society/Edith Blake)

residential America the family home is well insulated from the public street by a barricade of open lawn, if not fences and plantings, while public space of a generous nature is almost never integrated into the community fabric, much less climaxed by a working monument of aesthetic quality. Wesleyan Grove, almost alone, exaggerated and made magnificent the spatial characteristics of the nineteenth-century Methodist campground, an institution with its own set of social and religious goals backed by a tradition of environmental art that was geared to achieve them. Anonymous planners and builders at the campground invented the forms, in cottage and layout, to house a community of crowds, residents, and mobs of curious day-trippers, achieving an urbanlike density of people and buildings hidden in untrammeled nature, "a city in the woods." They took the pervasive American event, the camp-meeting revival in the woods, and brought it to spectacular conclusion in a form which has someting to do with that pervasive American

Cottages in campground, 1870s. (Dukes County Historical Society/Edith Blake)

residential habit, the suburb. Both, after all, were intended as societies of the like-minded, with strong family ideology, living in nature.

In the 1860s, with Wesleyan Grove attracting the curious as well as the religious, canny real estate entrepreneurs purchased the land adjacent to the campground for a summer resort. Oak Bluffs, an early example of a designed subdivision, was the first and most direct application of the lessons in place-making learned from the famous camp meeting. These developers consciously extended the spirit and forms of Wesleyan Grove, using professionals—a landscape gardener and architect—to replicate the compelling magic, the otherworldliness of the original. They, together with still more

Cottages in campground, 1980s. (Ellen Weiss)

anonymous builders, succeeded in their goal, adding a towered skyline to make explicit the implied urbanity of festive throngs of religious seekers.

The growth of Wesleyan Grove from a few tents in the wilderness to the famed city of cottages in little more than 50 years, and of Oak Bluffs, the instant "fairyland" next door, was well observed by anonymous correspondents to city newspapers. They recorded not only the emergence of built form, but, more important, the moods and some of the motives behind it, including disagreement over growth, materialism, and secularization. It is through these correspondents that we can grasp the events behind the physical phenomena, knowing in the end that the environment which we see today or reconstruct in our minds replicates in form and space the senses and spirit of a devout and lively people. Wesleyan Grove and Oak Bluffs, now monuments, were once the scaffolds of an extraordinary human experience.

H A V E N
Crystal L.
New York and Portland Steamers
Old Camp Meeting L'd'g.
EASTVILLE
VINEYARD HIGHLANDS
Lake Anthony
Oak Bluffs L'd'g.
Sunset L.
COTTAGE CITY
Lover's Rk.
LAGOON HEIGHTS
Farm Pond

CITY IN THE WOODS

ONE

CAMP MEETINGS

THE RELIGIOUS CAMP MEETING, an American phenomenon, was an open-air revival lasting several days, during which the participants had to sleep at the revival site because they were far from home. It went from invention around 1800, probably in Logan County in southwestern Kentucky, to fame a year later with the notorious meeting at Cane Ridge in Bourbon County, central Kentucky. Within a few years, camp meetings were being held all over the western frontier, the South, and even in New England, spreading with amazing rapidity. The first meetings were the work of Presbyterians, with Baptists and Methodists taking part. But the first two denominations soon abandoned this arena of religious labor, leaving it to the Methodists. This group embraced the camp meeting as an exercise and, simultaneously, grew to be the largest single Protestant denomination in the country. As camp meetings spread across the country, listed Methodists increased from 2,800 in 1800 to 30,741 in 1812, and to 1,068,525 in 1844.[1]

Much of the popularity of camp meetings was due to the devotion of the itinerant Methodist preachers, unschooled frontier circuit riders, many of whom had themselves been led from a sinful state to the religious life at a forest revival. Bishop Asbury, head of American Methodism from the Revolution until his death in 1816, saw these "wooden Bethels" as devices of great strategic importance, and often urged their usage. This sanction by a higher authority was important, for emotional excesses made camp meetings controversial within the denomination, and they never received sanction by the general conference. Unlike class meetings, societies, conferences, and other units of the highly developed denominational structure, camp meetings were never mentioned by Methodism's manual and final arbiter, the *Discipline.* No records of their existence were kept by any central body, so that their time, place, and numbers of participants or conversions are known today only from occasional mentions in local histories or preachers' journals. Modern historians have estimated that by 1810 most circuits were participating in an annual camp meeting; 600 may have been held in 1816 and 1,000 in 1820.[2]

It was the Cane Ridge meeting of August 1801 that may have created the wave of forest revivals which fanned out from central Kentucky throughout the western frontier and back to the East Coast, and which precipitated so much controversy. Cane Ridge was the most disorderly, and thus the most famous, of all the early camp meetings, but it was also probably the least typical because of its extremes. It lasted six frenzied days and nights and involved 10,000 to 25,000 participants at a time when nearby Lexington, the largest town in the region, had only 2,000 inhabitants. The meeting was said to have made a noise "like Niagara" which was heard for miles, and people fell by the hundreds "as men slain in battle." The falling exercise was the most common manifestation of religious experience but other activities—barking, jerking, dancing, running, and speaking in tongues—were also observed. The English Methodists rejected the camp meeting as a religious tool partially because of the fame of Cane Ridge and these kinds of demonstrations.[3]

At a typical frontier camp meeting, the audience sat in a semicircle of backless benches facing the preacher's stand. The audience was divided by sex, in accordance with John Wesley's directive for regular services. Preaching emphasized the universality of human depravity contrasted with the hope of salvation by grace. A listener realizing his sinful state (thus "convicted") came down from the audience to a fenced area known as the "altar" or "pen" just below the stand. There he would be helped by the preacher, who might descend from the stand, and by lay exhorters, who could be women or even children. Exhorters would stay with the penitent, praying and singing with him until conversion—the experience of God's grace—relieved his despair. After conversion, the probationary Methodist would be assigned to a class, a small religious study group near his home, and be

Unidentified camp meeting. (Rev. B. W. Gorham, *Camp Meeting Manual,* 1854)

trained for church or "society" membership. For those already converted, camp meetings were opportunities for renewal, for mending a "backslidden" state, and for further spiritual development, often progressing toward the goal of Holiness, Sanctification, or Perfect Love, a state of sinlessness or perfection promoted by John Wesley and promulgated by many Methodists until the last decades of the nineteenth century.[4]

The long-term value of conversions made in camp meetings was debated throughout the first half of the nineteenth century. (One observer thought that because camp meetings were not mentioned in the *Discipline,* it was the only Methodist concern that could be argued freely, without fear of censure.[5]) Nathan Bangs, an historian of American Methodism writing in 1839, was an ardent defender. It is indisputable, he argued, that these meetings occur, that hundreds of people are prostrated, crying for mercy for their sins, and that many of these led afterwards "peaceable lives, in all godliness and honesty." He admitted that in the circumstances of such peculiar excitement there might be some disorder and "some mingling of human passions not sanctified by grace." But the biblical conversions of Daniel and Saul showed precedence for such convulsive experiences, as did the mid-eighteenth-century revivals of Jonathan Edwards and John Wesley. Since the mind and body are intimately connected, Bangs argued, the despair of the awakened sinner or the joy of the newly saved could surely affect the body in the extreme fashions seen at camp meetings. Normal human functions could be suspended by the "powerful operations of the Spirit of God," as men are convicted of sin or their souls filled with pure love. Man is a creature of passions as well as of intellect, and Christianity seeks to regulate the passions, not to destroy them. Those preachers who addressed the understanding only, as if men were merely intellectual beings, "avail themselves of not one half of the motives with which the gospel furnishes its servants."[6] And if the person who has professed penitence and experienced conversion goes on to lead a righteous life, then one must conclude that the work was effected by the Spirit of God.

The experiential source of Bangs's reasoned defense is revealed in a description of a camp meeting he had attended 21 years earlier. In 1818 he was 40 years old, with 16 years of Methodist preaching behind him, and was chairman of the committee charged with expanding Jesse Lee's 1810 history of American Methodism. Bangs's description of the meeting at Cowharbor, Long Island, began, as so many do, with an account of travel to and nature of the grounds. Between 40 and 50 sloops arrived at the waterside site, bearing 6,000 to 8,000 people. Great order and solemnity prevailed amid the beauty of the scene. The gentle zephyrs softly whispering through the foliage of the beautiful grove, now consecrated to God, were an expression of that divine Spirit which so sweetly filled the soul and tranquilized all passions of the human heart. No turbulent passion was permitted to interrupt the sacred peace and divine harmony. The exercises were solemn and impressive and the "falling tear from many eyes" witnessed the inward anguish produced. About midnight Bangs was attracted by the shouts of a close friend who was among those who had been overwhelmed. Going to his aid, he at first looked on with the criticizing eye of cool philosophy, determined not to be carried away by passionate exclamations. Bracing himself as much as possible, he was resolved passions should not get the ascendancy over judgment. "But in spite of all my philosophy, my prejudice, and my resistance, my heart suddenly melted like wax before the fire, and my nerves seemed in a moment relaxed." The showers of grace descended, along with some real rain, and Bangs joined those others, happy and sleepless, who sang the night away:

> With thee all night I mean to stay,
> and wrestle till the break of day.[7]

The following morning the sermons were pointed, lively, and solemn. The prayers were ardent, faithful, and persevering. The singing was melodious, and calculated to elevate the mind to the third heaven. The shouts of redeeming love were solemnly delightful; and the cries of penitent sinners deep and piercing. Even though all had labored incessantly the previous 24 hours, many continued again deep into the night, "their souls being knit together by divine love." On the last day of the revival, the exercises were again marked by the mournful cries of penitent sinners and the ardent prayers of the saved. Those who were to go to New York had to remain the night, so services were continued still another evening, with many falling to the ground.

> This was one of the most awfully solemn scenes my eyes ever beheld. Such a sense of the ineffable Majesty rested upon my soul, that I was lost in astonishment, wonder, and profound adoration. Human language cannot express the solemn, the delightful, the deep and joyful sensation which pervaded my soul. Nor me alone. It was a general shower of divine love. It seemed as if the windows of heaven were opened, and such a blessing poured out that there was scarce room to contain it.

Song, prayer, and exhortation continued until three in the morning. At eight, Bangs was on the boat headed for the city, "going into the world again," but his meditations continued.

> The writer of this imperfect sketch feels as if he should praise God in eternity for this camp-meeting. . . . It is not a transient blaze or a sudden ecstasy. . . . Sometimes when I have indulged in the cool speculation which worldly prudence would suggest, so many objects have been raised in mind against camp-meetings, that I have been ready to proclaim war against them; but these objections have uniformly been obviated by witnessing the beneficial effects of the meetings while attending them. My theories have all been torn in pieces while testing them by actual experiment. . . . What I experience I know.[8]

Bangs did not fear the Cane Ridge type of excesses although he did urge order and control. He seemed more worried that without good planning, camp meetings might degenerate into seasons of idle recreation. The charge of "pic-nic spirit" rather than emotional extremes was in the air in the decade following Bangs's history, and by 1854 the second major defender of the camp meeting as a religious form addressed the new issue extensively. The Reverend B. W. Gorham was a New York and New England preacher who wrote a slim volume, the *Camp Meeting Manual,* which was both a practical guide to campground layout and a moving account of conversion and sanctification experiences. Gorham addressed the growing recreational aspect of revivals by arguing from Old Testament analogies, frankly embracing extra-religious pleasures rather than denying them. The Hebrew people, he argued, achieved communal joy, national strength, and intensified religious purpose through holidays. Social bonding was necessary to a dispersed and persecuted people. The September festival of Sukkot, the Feast of Tabernacles, was especially important because it fulfilled the biblical (Lev. 23:43) injunction "Ye shall dwell in booths [tents] seven days, all that are Israelites born shall dwell in booths." Gorham had proposed a translation of "booths" to "tents," thus getting God's instruction for a religious convocation "identical with a modern Camp Meeting."[9]

Other useful Old Testament justifications of wilderness festivals are in Deut. 16:14–15, which urged including servants, orphans, and widows in the seven-day feast, and Neh. 8:14–17, which instructs readers in building booths out of branches—possible source for the arbors of southern camp meetings. All nations, ancient and modern, continued Gorham, hold great political and religious festivals, proving that something in man demands these excitements.

Hebraic analogies for camp meetings appeared in other Methodist writings. In 1839 the *Western Christian Advocate* likened revivals to the Feast of Tabernacles as a reminder to the Jewish people "of travels of ancestors through the wilderness, when they lived in tents . . . pilgrims with no certain dwelling place so that they must seek another and better country. . . . By repairing to the grove and spreading our tents we declare in actions louder than words that this world is not our home and that we seek a city whose builder and maker is God."[10] The sense of the sacred "other" place, solace for the insecure, permeating the quotation has Utopian ramifications in one direction and, in another, absurdly, the resort ends that so many camp meetings eventually achieved. "Canaan" and "Beulah Land" frequently appeared in the literature in reference to the campground itself, the Old Testament name for the promised land heightening the intensity of feeling attached to the site. As late as the 1870s Wesleyan Grove had a bridge over a pond that was called "the crossing over Jordan."

Gorham used Hebraic history for other points in his camp-meeting defense. The ancient Israelites continued their Feast of the Tabernacles as a religious meeting away from their homes, even when they were a settled people with synagogues in every village and a temple at Jerusalem which could accommodate thousands. This answered the argument that camp meetings were all right when Methodism was new and its adherents poor and dispersed, but now that the well-to-do population was served by large churches, a device "fitted only to a rude society" should be abandoned. Revivals for a settled and substantial people served other functions. It was good to take people away from the worldliness of their homes and the pressures of business and to get ministers together to promote Christian union. The effect of meetings on ministers is given with poignant specificity: They must be brought together to neutralize their isolation and egomania, so that they would "lose selfishness and pride and unholy ambition while bathing together in a common ocean of love."[11] Gorham himself may have had just the problems for which he had prescribed the camp-meeting cure. He was characterized in the Massachusetts press as a "camp meeting to camp meeting runner" who talked too long and had a dictatorial manner. On one occasion a commentator was relieved that Brother Gorham and his uncouth songs were not present because he was "entertaining" at another meeting.[12] Whatever his personal effect, we must be grateful to him for furnishing us with the finest justification of camp meetings we are ever likely to have: "The truth is, human life needs to be dotted over with occasions of stirring interest. The journey asks its milestones, or rather, if you please, its watering places along the way. Our nature requires the recurrence now and then of some event of special interest; something that shall peer up from the dead level of existence—an object for hope to rest upon in the future—an oasis in the desert of the remembered past."[13]

To know the physical form of camp-meeting grounds, the setting of the religious events so eloquently defended by Bangs and Gorham, one must consult a variety of sources—one or two instructional manuals and the accounts of travelers, preachers, and participants. As early as 1810, Jesse Lee included guidelines for organizing camp meetings in his brief history of American Methodism. He suggested an "oblong square" of tents on 2 to 4 acres of land which had been cleared of undergrowth.[14] Behind the tents should be an area for carriages and wagons, and behind that another area for horses. Fires for cooking could go in back of the tents, or in front if one wanted to increase lighting for nighttime services. Lee suggested either one or two preachers' stands and candles at night to increase the mood of solemnity. There must be a guard and an announced order for each event. Lee's numbered list of instructions is in the style of the *Discipline,* suggesting that he was intentionally filling the void left by the Methodist handbook.

Early descriptions reveal a variety of forms for campground layout. Nathan Bangs described grounds in a circular form with rows of up to 200 tents "from three to six deep, and arranged on several streets, numbered and labelled, so that they may be distinguished one from another, and passed between."[15] Cooking fires were usually behind the tents to keep the smoke away from the meeting. There were lamps to illuminate the entire grounds at night and a mandatory light inside each tent all night long, presumably to lessen disorderly conduct. Western circuit rider James Quinn described a meeting with two preaching stands but no altar. Preaching alternated at the two stands throughout the day, but only one was illuminated at night. The *Western Christian Advocate* in 1839 recommended a campground plan that is reminiscent of Bangs's—an aisle around the first row of tents, fires outside the aisle, wagons and carriages outside of that, and horses beyond it all. "Then order, beauty and convenience of the whole would be such as becomes a place of worship." A few years earlier this journal had suggested that the square or circle of tents should be like the walls of a meetinghouse so that everyone coming there would behave as if they were in a church. "Those who would walk, talk, smoke or laugh would have to do it outside."[16]

Benjamin Latrobe, America's first professionally trained architect, provided the best record of an early campground with his sketch plan and two sections of a site that he saw in Virginia in 1809. Men and women were seated separately. Their areas were bounded at the rear by two semicircles of tents, one row behind the other, so that the audience, facing the preachers' stand, was in a loosely classical auditorium form. The two rows contained about 150 tents with fires for cooking in front. A separate row of about 30 tents for the Negroes was ranged behind the preachers' stand, completing the grounds by making the straight stem of a "D." The encampment was tightly enclosed by a wattle fence, leaving wagons and horses outside.[17]

Campgrounds seem to have been sanctified in a formal sense, as judged

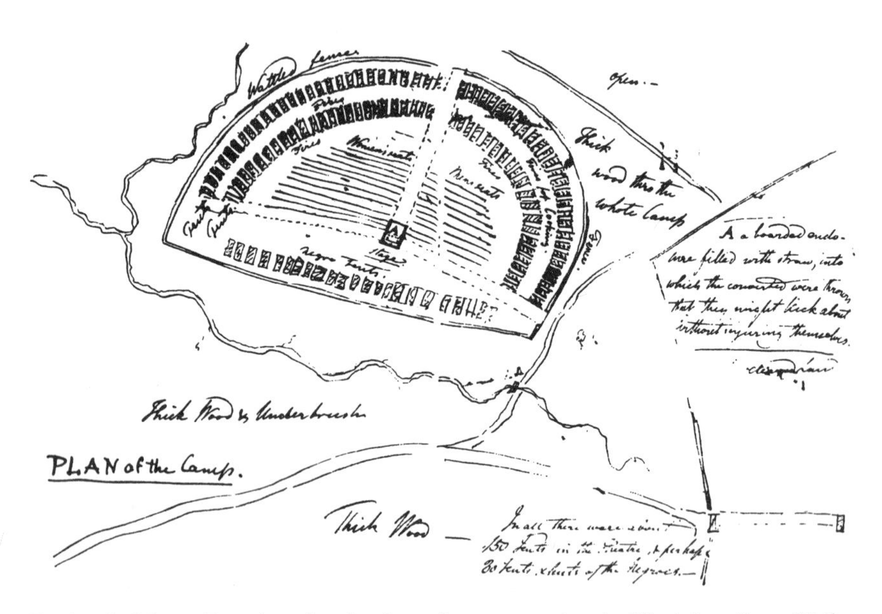

Benjamin Henry Latrobe, sketch plan of camp meeting in Virginia, 1809. (Talbot Hamlin, *Benjamin Henry Latrobe,* 1955)

from the many allusions to "consecrated grounds." Records of an actual dedication have not come down, however. Methodists had a habit of consecrating religious structures, and mid-nineteenth-century editions of the *Discipline* provided dedicatory services for cornerstones and completed churches. Twentieth-century editions expanded the tradition with services for organs, schools, hospitals, and nursing homes. There are examples in which the sacred ground of an early camp meeting was preserved and reused by being turned into a cemetery for those who found spiritual birth on the spot.[18] J. B. Jackson, this country's most provocative cultural geographer, felt that campgrounds were sacred in accordance with a long Judaeo-Christian tradition of defining sites as significant because of human or divine action, as opposed to the Greek sense of the sacred place as preexisting by definition of nature. Jackson was struck by the lack of scenic drama, the mundane quality of the landscape at the campgrounds he visited, and he seemed surprised that meetings were not set on the most dramatic sites available. What he did not understand was that shelter, privacy, and isolation were requirements for revival sites, not openness, drama, and splendid views.[19]

Even if the landscapes of revival sites were ordinary, travelers' accounts tell us that firelight, moonlight, and flickering candles turned campgrounds into scenes of great emotional power at night. Frances Trollope found a night meeting "solemn and beautiful" before the religious effects took hold. Then the sublimity of the occasion was lost as she found herself horrified by what seemed to be human degradation, lascivious male exhorters attending hysterical young girls.[20] Another traveler described a night meeting in terms of sheer exaltation, "the fires reflecting light amidst the branches of the forest-trees; the candles and lamps illuminating the ground; hundreds moving to and fro with torches like Gideon's army." The sound of exhortation, singing, praying, and rejoicing rushing from various parts of the encampment, was enough to enlist the feelings of the heart and absorb all the powers of thought.[21] Lamps, pine torches, fireflies, and stars all mingled into another light-filled vision. For Thomas Low Nichols, a night revival scene elicited special attention. The camp was lighted by lanterns in the trees and blazing fires of pine-knots, the scene becoming strange and beautiful. Lights shone in the tents and gleamed in the forest, and rude, melodious Methodist hymns rang through the woods. There was a glittering phosphorescent gleam from roots rubbed raw by the feet of the crowd, a moon, and the melancholy scream of a loon across the lake. "In this wild and solemn night-scene, the voice of the preacher has double power, and the harvest of converts is increased."[22] "Firestands," or "fire altars," a common lighting device, increased the effect. These were wooden platforms 6 feet high, topped with packed earth and pine-knot bonfires that cast a bright, flickering light onto the moving leaves of trees above.

Camp meetings at sites which had to be approached by water offered another range of psychic dislocations to aid religious experience. The passage increased the distance from worldly matters, and the sound of lapping waters on shores close to the preaching area added to the joyous mixture of natural sensations. The 1818 Long Island revival which transformed the skeptical Nathan Bangs into a lifelong camp-meeting aficionado was by the sea, and Bangs later recommended that grounds be chosen "in such a place that people may go by water, in sloops or steam-boats."[23] Camp meetings were held at the shore at Eastham, near the tip of Cape Cod, from 1819 to 1858 and were especially beloved by the Bostonians who came by boat. Father Edward T. Taylor, the famed sailor preacher, was particularly attached to the Eastham meeting, as was his biographer, the abolitionist Reverend Gilbert Haven, later a resident of Wesleyan Grove.[24] The joys of the sea intensified the relaxation in nature and camp-meeting sociability, as can be read in a letter from Haven to his family.

> With talking going on as fast as it can, I suppose you will not expect a very connected discourse, but, after camp-meeting style, warm, hearty exhortation. I wish you were both here, taking the sea baths, hearing, seeing, loafing under these pleasant little trees, and having a quiet and delicious time. We had a delightful sail down. The waves were quiet and the moon glorious; the crowd was good natured, scattered around on the decks under the open sky, most of them without sleep. We arrived about four o'clock in the morning, and had a grand, dancing boat, leaping over the big waves on which we rolled to the shore, jumped off into the surf, and entered the nicest and quietest of groves. . . . It certainly is the most perfect spot for a camp-meeting I ever saw, with fresh sea air, magnificent bathing in the real Atlantic outside Cape Cod, with a grove of small but thickly studded oaks, and the barrenest sand hills with salt grass and scrub oaks.[25]

This was written early in the meeting of 1857, before the intensity of religious labor had taken hold. It records the relaxed and open ambience offered to the seeker, with coreligionists thrown close together in refreshing nature. Simple preaching facilities with, perhaps, some stunning nighttime light effects, were the only manmade intrusions at these early camp meetings. It would be the task of the ambitious revivals of the second half of the century to preserve this effervescent mix of sociability and nature, allowing the seeker to lose his anxiety and sense of isolated self and to receive the Christian blessing.

Early camp meetings were frequently said to have had the trees above as an "architecture." Manmade devices inserted within the natural scene were minimal, functioning only as tools that permitted action without interference with the ambient green so crucial to religious experience. The basic tools of the first part of the century were the preachers' stands, tents, altars, backless benches, and fire altars, all simple appliances not intended as aesthetic objects. By the 1830s, however, especially in the Ohio River valley and the South, new building types were being devised. These were large enough to impress themselves onto the scene and to begin to compete with the "architecture" of the trees. The most important of the new forms were wooden cabins, to shelter worshipers at night, and tabernacles—great open sheds that covered the entire audience during the revival itself.

By the end of the 1830s substantial wooden cabins or cottages probably existed. In 1838 a campground near Cincinnati was described as having not only hundreds of cloth tents, "their snowy whiteness contrasting beautifully with the deep verdure and gloom of the forest," but also board cabins, some of which were two stories high.[26] A nearby meeting of 1840 was a "village in the woods," with two-story frame buildings with dining and prayer meetings on the ground floor, and lodging and private devotions on the second. These cottages might have been like those of the Duck Creek camp meet-

Duck Creek camp meeting, near Cincinnati, 1852. (*Gleason's Pictorial*)

ing, then 5 miles east of Cincinnati, shown in *Gleason's Pictorial* in 1852. The audience and preaching stand are surrounded by a veritable wall of two-story wooden buildings with shed roofs, the high edge of the roof on the inner or audience side. These odd structures were made of horizontal planks with one or two of the boards left out on the second story, forming an opening the width of the entire building from which about eight or ten people are shown watching the camp meeting in progress below, as if from a theater balcony. No other openings are visible from the audience side. With the "balcony" slit higher than the preachers' stand, the buildings have an ominous bunkerlike appearance, perhaps an interpretation of the biblical "booths" of the Israelites. Great trees, which loomed in turn over the cottages, dwarfed them within nature.[27]

Southern camp meetings also built wooden cabins (called "tents,") at an early date. Southern versions are usually end-gable one-story structures of an appealing simplicity in various arrangements and materials. Some of these primitive cabins may date from the 1840s and 1850s, but most that the author has seen are from the late nineteenth and early twentieth century, even though the revival site may be much older. Sawdust floors are common as are partial partitions to allow air circulation above and through screened ventilation slits in the exterior walls. The meeting at Salem, Georgia, dates to 1828 but has early twentieth-century "tents" which, like earlier ones at nearby Shingleroof, express community bonding with continuous shed porches, often with vine-covered arbors, running from cabin to cabin along the front. An architect in Mississippi has reported party-wall dogtrot* tents with the interior space above the partitions but under the gable continuous from one family unit to another, aiding ventilation. Community asserts itself again at the rear of the cabins, where the cooking and eating areas for each family are separated only by loosely spaced horizontal boarding.[28]

The most appealing building type in southern camp-grounds is the large tabernacle or "arbor" (or even "harbour"). These are great, roofed sheds, open on all sides and large enough to shelter the preachers' stand and the entire audience. Wooden tabernacles constitute an American building type of remarkable consistency over a wide geographic range—from Virginia to East Texas—and a wide temporal range as well. Of the 10 of these impressive forms recently listed in the National Register of Historic Places the earliest was built in 1832 at Rock Springs, North Carolina, and the latest in

*Dogtrots were a common Southern rural house form consisting of two rooms separated by an open ended, roofed breezeway, or dogtrot. Most party-wall houses are row houses, sharing side walls, and are found in cities, so that party-wall dogtrots, joined by side walls, are an anomaly.

Continuous porches connecting four "tents" at Shingleroof campground, near Atlanta, Georgia. (Ellen Weiss)

1942 at Ebenezer, Arkansas. The form may be almost as old as camp meetings themselves, for Peter Cartwright, writing of structures in Tennessee, Kentucky, and the Carolinas, referred to such sheds, including one that he said could shelter 5,000. Another source mentioned a shed built in 1807 at Goshen, Indiana, which was said to have covered 700.[29] All of the wooden tabernacles listed in the National Register are constructed of hand-hewn squared timbers, pegged or mortise-and-tenoned, with angle braces and exposed trusses. Some have several tiers of roof, with clerestory lighting and ventilation. Roofs are hipped or gabled, or both in combination, and were usually covered with wood shingles, later to be replaced by sheets of tin. Those southern wooden tabernacles are powerful architectural forms from a vernacular tradition.

For the physical form of pre-Civil War camp meetings in the Northeast, there is a published source, the Reverend B. W. Gorham's 1854 *Camp Meeting Manual,* that eloquent defense of the recreational spirit. Gorham preached in northern Pennsylvania, western New York, and New England, and the *Manual* represents the normal or best practice in this wide region.

(*Top*) Tabernacle at Little Texas, Macon County, Alabama, *c.* 1850. (Ellen Weiss)

(*Bottom*) Tabernacle interior at Little Texas. (Ellen Weiss)

This practice was based on the simple preachers' stand and cloth tents, not on cottages or large wooden tabernacles, a tradition different from the Ohio River valley and the South.[30] Gorham's *Manual* illustrated a northern meeting with a print shown on p. 5, which has been reproduced so often since that it has become emblematic of the revival form. Although the landscape and foliage seem to have nothing to do with the Vineyard, the signs on the society tents are suggestive of Wesleyan Grove, as is a copy of the print labeled "Martha's Vineyard, 1851," in the island historical society.[31]

Gorham opened his discussion of campground layout with a list of preliminaries. One must have a large district of Methodists in order to have an effective meeting; notice must be placed in the regional Methodist press several months in advance; a committee of arrangements must be appointed. The two best times to hold a meeting are June 20 to July 15, and August 20 to September 15, even though the first period is flawed: trees are then in a growth stage and tethered horses gnawing on their bark could do them harm. The meeting could last any length of time from five to eight days, but the time, once announced, should never be changed. When choosing a campground, one must look for good drinking water, a neighborhood sympathetic to Methodism, pasturage for the horses, shade for the audience, a location that is central to the district, a landowner who will allow wood to be cut for tent poles, a level or slightly inclined surface which is free from bumps, and a forest to screen the meeting from the world. If there were to be 100 family tents, or an equivalent in society tents, then a half acre of preaching space would be needed. The tents themselves would occupy another half acre. The grounds should be nearly circular, the space having been cleared of brush and the trees trimmed to a height of 10 or 12 feet. The preaching stand should be placed on the north side of the clearing, so the audience would face away from the sun.

Gorham's manual also deals with buildings and fixtures. The preachers' stand should be 6 feet off the ground, and an area 25 feet square is recommended for the altar. The altar should have entrances only at the corners nearest to the stand. It should have seats, and a level ground covered with sawdust or bark topped by a thick layer of straw (indicating that Gorham expected falling exercises). The central aisle of the main seating area should begin 35 feet from the stand, and this aisle should be 7 feet wide. The aisle would be located by the supports for the benches, these being 9 feet apart to allow for a 1-foot bench plank overhang and a 7-foot aisle. Secondary aisles could be narrower, but not less than 5 feet wide. "Places for retirement"—that is, latrines—must be designated, especially for the women.

Gorham provided a diagram of a meeting ground inscribed within an oval, the oval representing the front line of the tents.

Gorham also offered instructions for making family and society tents, together with a strong indictment of wooden cabins or cottages. Their appearance is "calculated to excite a class of low and ludicrous ideas, since they give the spectator rather the idea of a huddle of Irish rail-road shanties than of a worshiping people 'dwelling in the goodly tents of Jacob.' *Numb.* XXIV, 5."[32] Wooden tents are also dark and unpleasant within, not as waterproof as a properly constructed canvas tent, and more expensive. Gorham's advice for a family tent for six or eight mandates a 12-foot square clearing, a frame of three posts to a side and a sill to keep the cloth from the ground. Forty yards of "factory cloth" would be needed to cover the frame and provide a fly roof for rain-proofing. Cloth for society tents—those holding church groups—is to be sewn by the women in their homes, working together. Twenty by 30 feet in plan is the recommended size, and it must have a steeper pitch for the roof than the family tents. Gorham believed that even if the members of a church attending this meeting were all sleeping in family tents, there should still be a society tent for prayer

Family tents are still in use at Ocean Grove, New Jersey, a camp meeting founded in 1868 in imitation of Wesleyan Grove. (Ellen Weiss)

meetings. There could be commercial boarding tents if they were well controlled. Church groups might also hire a cook.

Gorham also recommended careful physical planning for other aspects of camp meetings. It was crucial that the meeting be well organized, for only by being a great event could it escape contempt and achieve its spiritual goals. As in the army, there must be a head, preferably the presiding elder of the district. There must be police with the authority to make arrests, for people take liberties at camp meetings that they would not take at home. Rules should be minimal but firmly enforced. A daily schedule should be posted and strong leadership should come from the stand. Sessions lasting all night at the end of the meeting are of doubtful utility. Preaching should be simple and direct, "not dry, dogmatic theorizing; not metaphysical hairsplitting; not pulpit bombast; but plain, clear, evangelical Bible truth, uttered with faithful, solemn, earnestness." With good preaching and careful management, the effective primitive camp meetings of the old days could be transmuted into a powerful tool for the new age, with its different spirit. For

> . . . beyond all precedent this is an age of progress and innovation in nearly every human pursuit. Old methods are abandoned, old opinions exploded, and old limits transcended on every hand. The hand wheel has given place to the jenney, the knitting needle to the loom, and the steam engine has knocked the shuttle from a thousand hands, to seize and propel itself with a thousand-fold velocity. The war-steamer's thundering trump is heard in every sea, and the rural villages are startled by the snort of the iron horse. The old county fair has suggested the World's Exhibition, and the circus has sprung into the hippodrome. In sin too men are unhesitating and rampant. They sin on a large scale, and with a bold hand.

Wickedness and depravity show a wonderful precocity. But Christianity is also strong. It is "the source of the immense impulses now driving human society forward through a larger number of stadia of progress in a decade than it once made in several centuries of years. Let her show herself equal to the times she has introduced."[32] The camp meeting is only one appliance the church has at her disposal to

> . . . strengthen her paternal bonds, nourish her life, and promote her aggression. . . . Let, then, the tribes of our Israel gather annually to the tented woodland. . . . Let every minister and every man, whether venerable in years, or fresh in youth, be at his post. Let there be no idlers among us, no preaching or praying for a name, no gossip, no ranting, no idle thronging.
>
> Then there would be no such thing as failure on the camp ground. The church would be quickened, the mourner comforted, and hundreds

awakened and converted. And dark-minded evil men, who would not yield to truth, should yet confess, "God is come into the camp."[33]

* ** *

ESTABLISHING METHODISM IN NEW ENGLAND was a frontierlike venture with its own problems. Bishop Francis Asbury, developer of American Methodism, found in 1791 that although he was never out of sight of a house and often in view of several steeples, the people seldom prayed and seemed spiritually dead. Jesse Lee, a young Virginia circuit rider in 1790, took it as a challenge to go into a region that was already settled and highly churched. Lee's first successes were in several Connecticut towns, where he managed to establish classes, and at Lynn, north of Boston.[34]

Building on the preliminary work of a local man who had heard Methodist preaching in the south, Lee converted the majority of the Lynn Congregational church, causing anti-Methodist diatribes in the town press as late as 1855. The arduous lives of western frontier circuit riders had some echo in New England: daily preaching, traveling on horseback in bad weather, risking ill health, being homeless, and depending on others for food and lodging, all for virtually no salary. New England circuit riders also suffered from scurrilous attacks in the press and were stoned or set upon by dogs for representing the alien doctrine. Often only the poor would help, it seemed, causing circuit preacher Elijah Hedding the added pain of knowing that he might be accepting food from people who needed it perhaps more desperately than he. On occasion, even decent townspeople who might be counted on for acts of kindness in other situations were put off by the preacher's lack of education. A New England lawyer once tested Jesse Lee by questioning him in Latin before a group of onlookers. Knowing no Latin, Lee said something back in German which was taken by the lawyer as Hebrew, permitting the quick-witted preacher to pass as "learned." Others were less fortunate.

Active persecution of Methodist groups was also a sorry fact of New England life. In Provincetown in 1795, the town voted that the 30 Methodists should not be allowed to build a meetinghouse. This was taken as a mandate to tear down an existing building and to tar and feather a Methodist in effigy. The beleaguered Wesleyans survived, built and protected a second church (attending it through fishhead peltings), and eventually gained numerical strength to vote the Congregationists out of their meetinghouse. By 1860, the Methodists had a splendid edifice with a 162-foot tower and seats for 900. In general, in New England, the denomination did better along the coast. New London, Connecticut, reported 50 members in

1793. Warren, Rhode Island, had a Methodist meetinghouse as early as 1794, and Nantucket had 65 members and a new building in 1800.

The camp meeting came quickly with the spread of Methodism in New England. A "boatload of saints and sinners" sailed down the river from Middletown, Connecticut, to a revival at Haddam in 1802, only a year after Cane Ridge.[35] A great revival in 1809 in Monmouth, Maine, led to a massive religious outpouring in that region. Elijah Hedding, presiding elder during the Monmouth meeting, organized a meeting at Hebron, Connecticut the following year to which participants came from a 60-mile radius, provisioned for a week. Five hundred people were felled during one evening sermon, and many more, coming from nearby villages to see the phenomenon of 500 lying immobilized, were in turn converted. Memorable camp meetings continued through subsequent decades all through New England. The famed "Camp Meeting John Allen," a "bold, blatant, rum drinking, fun making Universalist" in his youth, was converted at a meeting at Industry, Maine, in 1825. A figure at all New England revivals, he finally found Perfect Love at the first National Camp Meeting Association Holiness meeting in 1867. He died at 92 having attended 374 revivals.[36]

Methodism on the Massachusetts island of Martha's Vineyard followed the general New England pattern: sporadic preaching by laymen or circuit riders, recorded in histories but ineffective in terms of building a church; a longer period in which classes and then societies were formed, a period marked by persecution; and a final triumph monumentalized in architecture. In 1787 two former Virginia slaves, who were Methodist lay speakers, arrived on the island and preached for several years. Jesse Lee visited, as did other circuit riders, without much long-term effect on the vice-ridden island whaling port of Edgartown, according to Jeremiah Pease, historian of Vineyard Methodism.[37]

Permanent work dated from the arrival of Brother Erastus Otis in 1809, and enough converts gained to attract persecution. "What indignant feelings would arise in your minds, my friends, if while seated here your dresses should be besmeared with sand and mud, a dead cat thrown into your midst and your ears saluted with the hideous howling of a large dog?"[38] A town petition was got up to drive the Methodist preachers away for using incantations, calling upon children and weak-minded persons, threatening them with damnation "as if they had the great power of God," for preaching blasphemy and for threatening the stability of the town. Brother Otis said that he often feared for his life during this time and felt it prudent to hold meetings in the countryside, rather than in town. Jeremiah Pease was, at this stage, among the scoffers.

The great Methodist advances came in the early 1820s with the arrival every year or two of a different but always remarkable preacher. The brief tenure of "Reformation" John Adams, so known because revivals seemed to follow him to all of his posts, saw the conversion of Jeremiah Pease and his family, leaders of the Congregational church and of the community.[39] Pease's life thenceforth would be filled with activities of his new commitment. His diary records class and prayer meetings, travels to the Cape or New Bedford for religious work, and deathbed watches for coreligionists—all noted together with jottings about the weather, arrival and departure of ships, and his varied professional activities.[40] Pease was a lighthouse keeper, customs collector, bone setter, shoemaker, and surveyor. He was also one of the founders of the Wesleyan Grove camp meeting, the traveling class leader who selected the site during one of his trips to the hamlet of Eastville.

Another early conversion which would be of special importance to this story was that of Hebron Vincent, one of 11 children of an impoverished Vineyard farmer and his devout wife.[41] Apprenticed at 13 to a shoemaker, this remarkable man became a self-educated educator, minister, attorney, and historian. In this last capacity he wrote, in two volumes and some additional papers, the only substantial history of any American camp meeting. He attended the first meeting of 1835 and was its secretary for at least two decades. His understanding forms our understanding of the events more than a century later.

The growth of Methodism on Martha's Vineyard from the 1820s was strong and consistent. Jeremiah Pease measured it by the succession of ever-larger meeting houses. In 1828, after earlier assemblies in private rooms and in a small building held jointly with the Baptists, the Methodists built an austere Federal-style structure, 50 by 40 feet in plan, without tower or steeple, in accordance with strictures in the *Discipline.* (This building now serves as town offices and a theater.) In the same year the Congregationalists also erected a new church, a more ambitious one than the Methodists', with a tower and steeple and elegant detailing from an Asher Benjamin pattern book, but 15 years later the Methodists built an even larger sanctuary, the famed "whaling church," now Edgartown's monument. Consecrated in 1844, this impressive Greek Revival structure, 75 by 62 feet in plan, had a freestanding hexastyle Doric portico and a two-stage tower with a Gothic ogive opening. The Congregationalists never expanded beyond their 1828 building. By the 1850s, with handsome new Greek Revival churches in Chilmark, Lambert's Cove, and Vineyard Haven, Methodism was the island's largest denomination.

The conquest of Martha's Vineyard by Methodism may indicate some special aspects of island life and may also illustrate an appeal of the religion's liberal doctrines and warmhearted emotionalism to the seafarers of all of coastal New England. The Reverend Samuel Devens, a Unitarian who visited the Vineyard in the 1830s, had noted a tendency to frequent and fervent religious outpourings, which he attributed to a moody temperament among islanders due to anxiety for men on whaling voyages, too much coffee, or an inherited nervousness intensified by intermarriage in an isolated community.[42] Whatever the reason, the taste for religious excitement produced, in 1853, a protracted winter meeting lasting 100 days that was so remarkable it was described at length in a history of southern New England Methodism at the end of the century. Apparently this revival converted just about everyone, including an intellectual whose experience, later described by himself as being like Saul's, stayed with him through a subsequent distinguished legal career, including a period as U.S. Consul in Canada.[43] Charles A. Johnson's theory that a theologically optimistic creed, whose truths were to be experienced directly, was needed to counteract not only the loneliness and deprivation of frontier life, but also its real dangers, may be useful for understanding its power in settled and churched areas.[44] Nearly every Vineyard family had members on a lengthy Pacific whaling voyage. Long separations from home and loved ones, extremes of boredom, discomfort, excitement, disease, and sudden violent deaths were the familiar lot of whalemen. Father Edward T. Taylor, the "sailor preacher" praised by Channing, Emerson, Melville, and Dickens, and founder of the Mariner's Bethel in Boston, held emotionally charged shipboard services in Edgartown for departing seamen and their families early in his ministerial career. Methodism in New Bedford was begun by sailors who had heard preaching in England. And although the famed Seaman's Bethel of New Bedford was interdenominational, it was the Methodists who led in its formation, as well as in other attempts to bring law and order to the waterfront. Accounts of New Bedford's docks are reminiscent of "Rogue's Harbour," Peter Cartwright's lawless Logan County, Kentucky, before religious reform took hold. Alcoholism, immorality, and violence responded to Methodist social controls on the settled East Coast as well as in the frontier West.

TWO

WESLEYAN GROVE: THE FIRST GROWTH

RIGHT FROM THE BEGINNING, in the preface of his history of the Martha's Vineyard camp meeting, Hebron Vincent insisted that he was writing about a place as well as an event and an institution. This sense of place, one "hallowed by a thousand Christian associations," forever reverberating with ancient gospel sermons, earnest prayers, "pathetic appeals to the unconverted, and exhilarating songs of praise," never leaves his prose.[1] The place of one's birth and childhood, he argued, is always present in one's thoughts and feelings. If the salvation of the human soul is of such vast importance as it is believed to be by all Christendom, why should not the place and circumstances of its conversion to God merit a history? Hebron Vincent offered his to those whose spiritual birth was on Martha's Vineyard.

Jeremiah Pease, leading Edgartown citizen, chose a site for a regional camp meeting, one to serve the Vineyard, Nantucket, and the adjacent areas of Cape Cod and southeastern Massachusetts, while traveling in the northern reaches of Edgartown township in 1835. He selected a grove of large oaks close to Nantucket Sound, east of the road from the whaling port to the hamlet of Eastville on the Holmes Hole (now Vineyard Haven)

Wesleyan Grove. (H. Vincent, *History of the Wesleyan Grove Camp Meeting,* 1858)

harbor. The grove was on a gentle northwest-facing slope and was bordered by ponds to the north and open pasture to the south. Although close to the sea, the natural amphitheater faced away from the water, oriented toward itself in an introspective fashion. A half acre of ground was cleared of underbrush and a driftwood shed erected as a preachers' "tent," with a stand built onto its front to serve as a pulpit. In front of this was the usual arrangement of a temporary altar, consisting of a railing enclosing a space about 25 feet by 12 with benches to be used, mainly, by the singers during the preaching service, and as a place for penitent sinners to gather. Beyond the altar were backless board benches. Some of the nine society or church tents had sailcloth extensions to the rear, like awnings, to shelter members at mealtimes. A well was dug. "The waving trees, the whispering breeze, the pathetic appeals, the earnest prayers, and the songs of praise, as well as the trembling of sinners under the Word, and the shining countenances of Christians lighted up with holy joy, all conspired to say 'Surely the Lord is in this place.'" At the meeting's end, money was raised to purchase the lumber used for the benches and stand and the decision taken to keep this now-sacred spot as a permanent one for camp meetings. "Thus we lived, labored, and rejoiced," in primitive simplicity. And thus ended the "feast of the first year" in the newly consecrated temple.[2]

The glow of religious sweetness, nature, and "primitive simplicity" suffused Vincent's year-by-year account of the annual camp meeting for more than three decades. In time these themes were overlaid by the equally emphasized motif of expansion. But, at least in Vincent's opinion, constant growth and gradual institutionalization never overwhelmed the basic mix of nature, human warmth, singing, praying, and direct and simple evangelistic preaching. Success for the new meeting was aided by the fact that, for many, it was accessible only by water, as Nathan Bangs was to recommend a few years later. Island isolation not only increased the feeling of other-worldliness but also protected the meeting from populous towns with rowdy youths who would disrupt proceedings for sport. Edgartown's Isaiah Pease, Dukes County Sheriff, was a participant, and his mere presence was said to insure absolute order. The island location offered its own special beauty. The Vineyard had always seemed to have a kind of gentle loveliness, noted in accounts of Gosnold's voyage of 1602, accounts that have been credited as a source of the setting of Shakespeare's *Tempest.*[3] "But the place of the meeting,—who can properly describe it?" Such another spot could not be found on earth so nearly resembling Eden in its primeval beauty and loveliness.[4] Even for Vincent, living in a small port town with easy access to sea, sand, and cultivated and uncultivated village edges, an oak grove only 6 miles away could seem primitive and untouched. Personal immersion in the purity of God's creation would always be part of the power of this religious meeting, even in its fullest development many decades later.

Hebron Vincent's history of the camp meeting during its first decade and a half forces reconsideration of the generally accepted notion that camp-meeting fervor had waned by the end of the second decade of the century, leaving the events poised between controlled religious services and various recreational pursuits. As late as 1841, Vincent recorded a session in which worshipers who were not easily excited were shorn of their strength, and lay for hours without the power either to speak or move. On another occasion, "an awful sense of the presence of God pervaded the encampment, and the slain of the Lord lay upon every side." Later, all retired to tents to besiege the enemy in his lurking place. The battle waxed warmer and warmer until after ten o'clock, when the enemy gave way and the shout of triumph rang all through the lines. Some now retired to rest, but many remained upon the field to celebrate a glorious victory. Prayer meetings in tents could go on deep into the night until exhaustion set in or the secretary of the meeting called round for the report: "We are almost all sanctified in our tent." "We have had a clean victory."[5] The military imagery so remarkable in these passages faded by the time of the Civil War.

Sometimes the vagaries of the weather could hasten spiritual progress.

On one occasion the congregation had already been on its knees for an hour when a violent storm came up. The flock remained in suppliant attitude before "Him in whom we trusted," through lightning, thunder, and a deluge of rain. It was a scene of moral sublimity which Hebron Vincent knew he would never forget. More often rain would only send the worshipers from the stand into the large tents for intimate prayer meetings, enclosed intensifications of the struggle for deliverance from sin and full salvation. The returning sun became, in turn, an omen of victory and joy, occasion to find salvation on every hand, and God's blessing as a special favor to this assembly.

But along with such vivid accounts of extraordinary religious experience Vincent could also, even in the 1840s, acclaim the meeting as a salubrious retreat from summer city ailments. After the cholera epidemic of 1849, the health-giving aspects of Wesleyan Grove elicited attention. The pure, exhilarating air under the shade of the tall oaks, the hearty greetings of old friends and acquaintances, the customary exercise of walking, and above all, the animating devotions of the occasion, enlivened spirits and sharpened appetites. The Christian invalid would realize better health by coming and living prudently a week, drinking of the living fountain of salvation that gushed up so freely and plentifully, than from a journey to the White Mountains, or a month's residence at Saratoga Springs. The Methodist camp meeting in which the "slain of the Lord" could lie for hours without the power to move or speak was becoming a health resort.

By 1850 Vincent was defending the meeting's recreational spirit, an anticipation of the Reverend Gorham's argument that communal joy strengthened religious purpose. "Tired nature occasionally seeks repose from the toil and strife of business. The ancient Jewish festivals were no less the means of restoring the social and intellectual equilibrium, than of promoting religious sentiment and devotional feeling." However, Vincent emphasized, it would be a foolish Christian who used only the natural and recreational aspects of the grove and failed to avail himself of the greatest spiritual opportunity of his life. One must be among those "happy hundreds" who left the cares of the world at home and came to this spiritual Jerusalem, this city of tents, to worship, not the leafy canopy which overshadows it, but the "Great Builder of this magnificent temple." The leafy canopy was the architecture but not the object of the event.[6]

While the camp meeting was developing as an emotionally laden religious experience interlaced with health and recreational benefits, it was also expanding. From nine society tents in 1835, the meeting grew to 12 tents in 1837 (plus a boarding tent for meals), to 16 in 1840, a number that may have included one or two small tents for families. The year 1840 also saw a step toward institutionalization: The meeting was given the name "Wesleyan Grove," a record book was purchased, and a five-year lease taken on the grounds. In naming their meeting, the leaders were following a Cape Cod revival called "Millenial Grove" held a year before—the first in New England to purchase land (1836) and incorporate (1838).[7] Wesleyan Grove continued a steady growth. In 1842 there were 1189 participants in 40 tents. More than twice that number of people were thought to be on the grounds for the Sabbath. In 1845 the meeting was moved to Westport Point on the mainland to keep it from becoming an "old story." The change was quickly rectified the following year, the Vineyard site reclaimed, and new seats and stand constructed to replace the 1835 equipment, which had been sold for lumber.[8] In 1849 a 10-year lease was taken. Religious sentiment was heightened that year because so many were leaving for the California goldfields.

Growth continued through the 1850s as well. In 1851 there were 100 tents, 3,500 to 4,000 Sabbath participants, and 134 conversions. A new seating arrangement with backs for the benches was now in place and Sirson P. Coffin, an Edgartown carpenter and lumber dealer who had been the agent or superintendent of the grounds since 1846, was praised for the work. The following year a committee of order was appointed to write rules of conduct. As amended in 1858, the rules set times for eating, sleeping, and services at the stand. Rules defined the area within the circle of society tents as sacred, located the latrines, and gave authority to society tent supervisors for everything relating to daily life. Rules also aimed at control of the increasingly numerous family tents, prescribing morning and evening devotions and requiring that a light be left burning all night in each tent, a traditional camp meeting regulation. In 1854 there were 36 society tents in the expanded main circle and a total of 180 tents of all types—society, family, and boarding. Vincent could now characterize the colony as a "snow white city." The year 1855 showed 150 family tents and 50 society and boarding tents. The following year, Vincent boasted of having preachers from Washington, D.C., Baltimore, and Ohio, and while the worshipers were still from the immediate southeastern New England area, soon Wesleyan Grove would have tenters and then cottagers from a wider northeastern range. In 1856 a tax was levied on family tents. There was discussion of moving the meeting because the oaks were dying, but new shade trees were planted instead. In 1857, the only year of which Vincent ever complained that the recreational spirit might outweigh the religious, there were 250 canvas homes. "It was delightful merely to behold this city of tents, the white coverings beautifully contrasting with the green foliage so gorgeously overshadowing them. But it was better still to mingle with the

Family tents, photographed in the 1870s. (Vineyard Vignettes)

population of this sequestered city, to listen to the word of the Lord preached by his faithful heralds."[9]

The expansion of the tenting grounds during the 1850s precluded the simple formation of a main circle enclosing the preaching space and led to a more complex plan than anything envisaged in Gorham's 1854 *Manual* or the squares and ovals of tradition. By 1857, tents had covered 12 to 15 acres and formed little neighborhoods with such fanciful names as "The Prairie," "Upham's Hill," and "Fourth Street Avenue." The latter was for inhabitants from the Fourth Street Methodist Church in New Bedford, who had stayed together as they moved from society tent into family tents. In 1857 a new lease was taken even though the old one was not up, securing the grounds until 1871. The winter of 1857–1858 saw a massive religious revival in cities across the country, with thousands attending interdenominational noontime prayer meetings. This awakening, combined with the new lease at Wesleyan Grove, according to Vincent, led to another expansion of the grounds in the summer of 1858: 70 new tents, some wooden buildings, and some tents with wooden walls. It was "quite a city, truly, and an exceedingly pleasant one."[10] Vincent praised the new wells and then, with typical interlocking of the material and spiritual, wished that participants might drink of both the healthful water of the grove and the "river of the water of life" to which they are so urgently invited by the men of God. Among the 12,000 present on the Sabbath were politicians—the Governor of Massachusetts and some bank commissioners—and members of other Protestant denominations. Even though only 6,000 or 7,000 could get close enough to hear the preacher, all behaved with decorum because they were imbued with religious reverence and a deep sense of the sanctity of the place.

Vincent's account of the year 1859 begins his second volume, the history of the meeting to 1869 and the record, according to the author, of a new era. In 1859 there were a few wooden cottages, new wells, and a large permanent building, the two-and-a-half-story camp-meeting headquarters, which still serves its original purpose. Placed outside the main circle of society tents, it included a room for baggage, a post office, an agent's office, and a lantern and oil room on the ground floor. The second story contained a lodging room for the agent and two meeting rooms, one of which could seat over a hundred ministers, tentmasters, and lay officials for business meetings. The third floor was used for visitors and storing tent covers over the winter. A Providence architect, Perez Mason, designed the building, although the porch which gives it so much character was built 11 years later, perhaps to Mason's design, perhaps not.[11] By 1859 there was also some attempt at organizing the grounds, which had been expanding in irregular fashion. The small family tents, which were behind the society tents, outside the preaching circle, were moved still further back, clearing a 40-foot-wide avenue around the big tents and the preaching area. The road was first called Asbury Avenue and later Broadway, and was meant to define a wagon route, keeping vehicles out of the preaching circle. Narrow lanes were cut into other densely tented areas to make them more "airy." At least one of these new lanes, Fisk Avenue, was placed to form a radial spoke from the main circle. By the following year there would be enough of these lanes to permit this "city in miniature" to be perceived as having a radial plan,

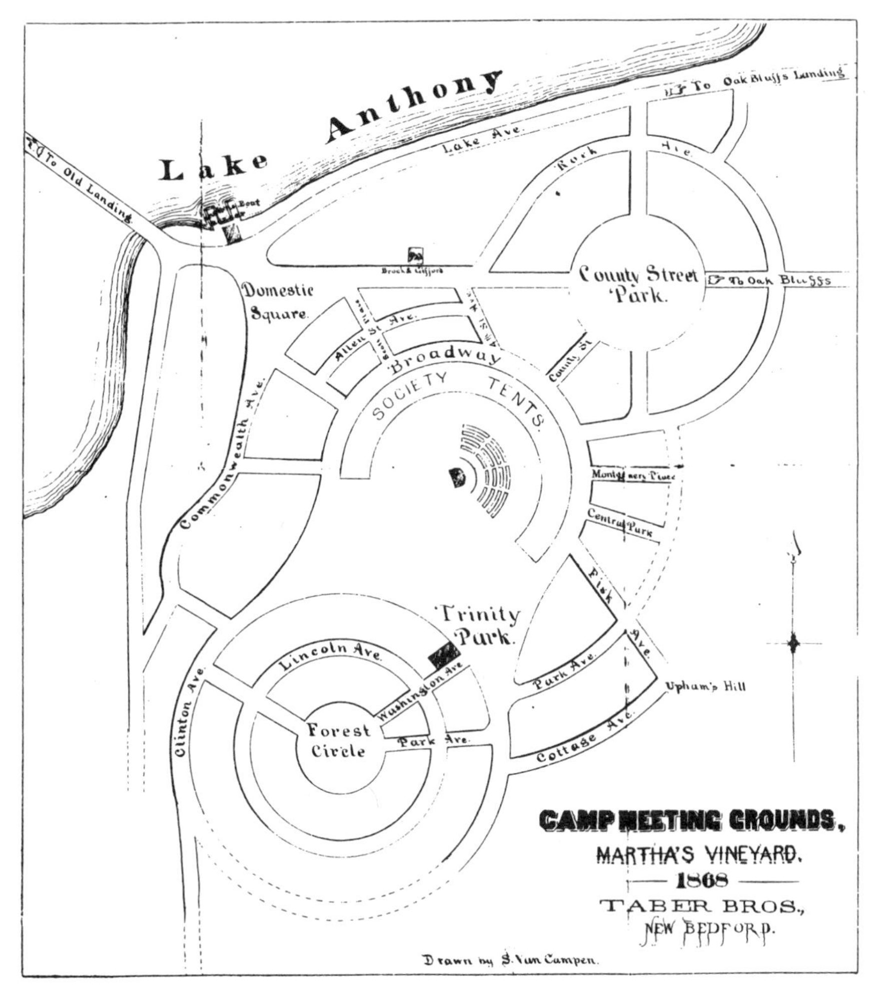

Schematic map of Wesleyan Grove, 1868. (Dukes County Historical Society)

"like a city in Germany, every street running to the center."[12] In 1860 there was also a second circular road outside of Broadway and concentric to it, serving two more lines of family tents. This second circular road is only partly visible on the plan of the grounds drawn in 1873 and even less so today.

The development at Wesleyan Grove of even a fragmentarily executed radial-concentric plan is of particular interest, for it is a plan type which, with a few notable exceptions such as Washington, D.C., and Detroit, was little used in America. The Vineyard radial-concentric form, evident by 1860, continued to be remarked throughout the decade. In 1867 *The New York Times* described Wesleyan Grove as having "circles intersected by avenues" and an 1868 tourist brochure included a diagram of the grounds which showed concentric circles cut through by radiating lanes, these juxtaposed with later tangential circles.[13] By the 1870s there were a number of campgrounds across the country with wheel plans, the central circles reserved for preaching and the radiating spokes for family tents or cottages. Since Gorham never suggested anything so complex in his 1854 *Manual,* it is possible that the form evolved on the Vineyard and was applied elsewhere, making the motif a testament to the primacy of Wesleyan Grove among post-Civil war permanent camp meetings. Round Lake campground (1869) near Troy, New York, had a radial plan and its founders had hired Sirson P. Coffin, Wesleyan Grove's superintendent, to help prepare the site. Gorham himself, a frequent Vineyard preacher, has been credited with the layout of a radial camp meeting at Manheim, Pennsylvania (1868). Pitman Grove, New Jersey (1870), has 12 spokelike streets radiating from its circular center, the number alluding, according to tradition, to the holy city of Revelation. Washington Grove, Maryland (1873), and Des Plaines, Illinois (1860 and 1865), also had radial or wheel plans, as did the camp-meeting grounds at Lancaster, Ohio (1878), and Plainville, Connecticut (1865).[14] The historical relationship among these examples of an unusual plan configuration must await analysis of each site. There is a possibility, if rather remote, of a "high-style" source for the idea of concentric circles cut by radiating roads, in the Circus Island section of a plan for Boston, published in 1844 by a Scottish designer, Robert Gourlay. If the Boston plan is the inspiration for Wesleyan Grove's radial-concentric parts, then the camp meeting is part of a grand European urbanistic tradition which includes the circus in John Nash's first plan for Regent's Park, London—a surprising if unlikely ancestor for a woodsy camp meeting. Wesleyan Grove's radial-concentric scheme can also be explained as the direct projection of the usual pattern of action and energy at a revival. It was most likely a vernacular landscape form which took on monumental dimensions, rather than a high-style idea which trickled down. It is, in landscape historian John R. Stilgoe's sense of the term, a common landscape.[15]

By 1860 the tented woodland meeting with its developing radial plan was passing through some critical point in its growth. Now it had a "village-like" appearance with "streets" of temporary homes. An extension of the main circle, a triangle called Coke's Park and later Trinity Park, was now the most aristocratic neighborhood with elaborate tents and the first artistically ambitious wooden cottage. In 1859 Vincent estimated the number of tents at 400, and a visitor from New York thought that this was the largest camp meeting in the nation. Yet all were impressed, and would be for decades to come, that, despite the mobs and the semi-urban density, the Edenic quality of the grove remained. "The beauty and loveliness of an evening view of the encampment, with its hundreds of tents lighted up, are seldom exceeded by anything in the world's panorama." The "hallowed and lovely spot, more beautiful than ever before," was now the annual resort of those who worshiped at the shrine of Nature or of God. Those who had left the abodes of men now entered the temples of nature "whose foundations were laid by the Almighty's hand, whose arches are the heavens and whose pillars are the morning light." Their whole being was penetrated by a feeling of deep calm, a repose that ended care, that sense of intimate companionship which was unknown in the city.[16] The quiet green, the whispering groves, the white tents, "like a flock resting in the shade at noontide heat," lulled the worshipers into a state of psychic suspension from which they could later be reconstituted as better individuals and regrouped into new social bondings.

Nature was still in domination, and could still admonish. One must learn from its beauty and teachings. "Let us be attentive listeners to the songs she sings in this beautiful temple. Let us trust more to our higher instincts, think purer thoughts, speak more truthful words, and then we shall fulfill our mission as do the birds and flowers and die happily."[17] Others, as well as Vincent, were now on hand to capture the spirit of the place. One elated city newspaper correspondent apologized for dwelling at great length on its beauties, saying that merely by lingering here for a while one would come away a better and wiser man. The environment alone could change one's life. Vincent judged the spiritual tone of 1859 higher than the year before. If there were only 30 conversions, they were of influential people of high social standing in their home cities. He reprinted an account by a Boston newspaperman to give a more removed view: This religious festival was to the Methodists like a Newport or Saratoga, without the great bustle, fashion, ceremony, and expense of such places, while the pure elements of religion were present as an additional luxury and enjoyment. Thousands roamed the grounds, many more than could be accommodated within the range of the preacher's voice from the stand, that sacred circle of Christians, within the great crowd, "from whence went up to the great God fervent prayers from the heart, and great and ever-living biblical truths." The Bostonian likened the Sabbath to the Fourth of July, but without fireworks and boisterousness. "Everybody appeared happy, joyful, smiling; thousands of young and beautiful females, elegantly dressed, promenaded along green paths with young men, whilst an immense crowd listened attentively to eloquent and deeply impressive addresses from clergymen, whose only object appeared to be to save souls."[18] Fashion and religion seemed in balance, and the creative tensions between the two would persist for decades in the life of Wesleyan Grove. And if there is some shading of opinion between Vincent and the newspaperman on the sacred versus secular, one senses that it was only because Vincent spent his time close to the stand, while the correspondent was at the more frivolous outer edges. The situation was rich enough in human possibilities to admit of both moods.

During the 1860s, Wesleyan Grove continued to develop in the manner established in the 1840s and 1850s: constant expansion fueled by a creative mix of heightened religious experience, human density, and nature. It became an institution. In 1860 Sirson P. Coffin proposed 11 articles of agreement for a Martha's Vineyard Camp-Meeting Association. Membership included the presiding elders of two Methodist districts from southeastern Massachusetts and Rhode Island, the ministers of participating churches, tent-masters of the society tents, and elected officers. The membership was to choose 15 Methodist laymen to serve as the finance committee, the association's executive arm. Coffin had been disturbed by a trend away from the primitive meetings of the 1840s, and the finance committee was meant to relieve the leadership of administrative burdens to concentrate on the spiritual. Tents and cottages now had to be licensed. In June 1865, four laymen operating as trustees of the camp meeting purchased the grounds, over 20 acres, for $1,300, a sum raised from tent and cottage owners. The Martha's Vineyard Camp-Meeting Association was incorporated by an act of the Massachusetts legislature in 1868. With its charter revised in the 1960s to make it a Protestant rather than a Methodist organization, this body today owns the land and orders life in the community.

The decade of the 1860s offered a new level of physical as well as institutional development in the campground. The years 1859–1860 produced still another preachers' stand, a seating arrangement for 4,000, and a system of whale-oil lanterns on posts throughout the grounds. The preachers' stand, which sat 30, was a well-thought-out structure with flat-arched lattice screens supporting a roof angled to act as a sounding board to magnify the speakers' voices. A rear wall made of movable shutters aided ventila-

Camp meeting with preachers' stand by Perez Mason (1859).
(*Harper's Weekly,* 1868)

tion. The designer was Perez Mason of Providence, the only professional architect with a continuing involvement with Wesleyan Grove. Mason designed a number of impressive Greek Revival and Victorian structures in southeastern New England, but only a handful of buildings at the camp meeting.[19] When the iron tabernacle was built in 1879 to the design of an engineer, Mason's stand was moved north "over Jordan"—demoted to a waiting room for passengers on the new horsecars. An identical preachers' stand remains in place at Asbury Grove, near Hamilton, north of Boston.

The grounds were greatly expanded during the decade of the 1860s and populated by an increasing number of a new architectural form, the campground cottage as well as tents. The method of grounds layout was an additive one of discrete neighborhood units, each built around small variously shaped parks, all scattered loosely on the land, as if without premeditation. In 1861 the encampment was described as a confusing jumble of 15 acres saved from disarray only by the clear order of the main circle of society tents and preaching space, the circle of Broadway around that, and the radial lanes. In 1863 the association removed some tent sites from the outlying but densely packed Upham's Hill to make Cottage Park a rectangular space, 200 by 30 feet, which quickly became a favored residential area. In 1864 the association made a neighborhood in "amphitheatral order," that is, on a radial concentric plan, mimicking the main circle. These 87 lots, initially called "Cottage Park" but within a year changed to "Forest

Circle," had a concentric outer road and an inner central park large enough for an ordinary camp meeting. About 70 feet in diameter, it would occasionally be used as an extra preaching space. Forest Circle is a powerfully articulated form, its clear and tight inner space a surprise to the pedestrian on a ramble about the grounds, and a strong and warmhearted imitation of the larger main circle.

The formal motif it established—small circle of tents and cottages in satellite or tangential relation to the enlarged radial-concentric main circle—was reiterated the following year, 1865, on the opposite side of the compound. The area that had been known as "The Prairie" was made into a circular park, slightly larger and less regular than Forest Circle (its diameter varying from 110 to 150 feet), not in "amphitheatral order."[20] The park was established too close to the main circle to permit a second enclosing concentric circle. The new location became County Street Park, named for the County Street Church of New Bedford. "The circular arrangements of the

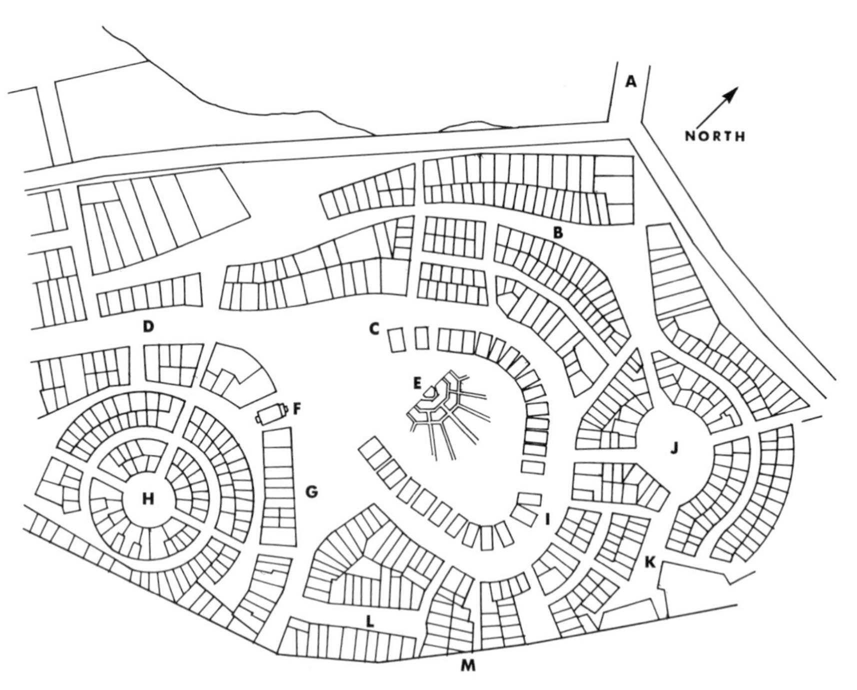

Lot plan of Wesleyan Grove in 1873. (Redrawn by Harold Raymond)
A. Crossing over Jordan; B. Commonwealth Avenue; C. Society tents; D. Clinton Avenue; E. Preachers' stand; F. Association building; G. Trinity Park; H. Forest Circle; I. Broadway; J. County Street Park; K. Montgomery Square; L. Cottage Avenue; M. Circuit Avenue

tents with their variegated furniture present a very picturesque appearance and as you stand in the center at evening and look around upon them all brilliantly lighted you almost imagine yourself within the charmed circle."[21] By 1865, so many individual neighborhoods had developed, each grouped about its own park, that Hebron Vincent could summarize the whole community in terms of these separate units. County Street Park was in the more northerly part of the ground, and was a delightful place. Near it was Fourth Street Avenue, much frequented. More central, and near the principal area, was Cottage Avenue, on Upham's Hill, which had all the charms of a little park, and was approached by Fisk and Park Avenues. Directly south of the main circle was Trinity Park, connecting with two entrances to the grounds, and with Broadway. A little south of this was Lincoln Park, a place of great comfort on a warm day. And still farther in the same direction was what was, for a while, called Cottage Park, now Forest Circle, being circular in shape, and surrounded by a regular system of avenues.

The grounds had yet to acquire Commonwealth Avenue, a long lane that is actually a segment of the lost outer concentric circle; Domestic and Montgomery Squares, triangles which were commercial zones in the 1870s, and which are now almost empty of buildings; Washington Park, a residential triangle south of Forest Circle; and, most important, Clinton Avenue. This last prestigious address, named perhaps for Clinton Avenue in Brooklyn, was created in 1868 by the closing of what had been the main entrance from the west. Clinton Avenue, a pedestrian boulevard with a wide median green strip bordered by macadam paths, loomed large in the accounts of the following decades for its grand cottages, its pride of place, and for its spirited entries into the mildly competitive concerts and illuminations provided by different neighborhoods for each other's entertainment.

The final effect of this ultimately 34 acre development is remarkable. Even today, without the revivals and the attendant throngs, and with the loss of many cottages and tents, the spatial definitions are clear, although set in a "mazy" overall scene. The main circle, now empty of society tents, is about 5 acres. In 1879 its central importance was fixed permanently by the construction of a vast iron tabernacle covering the preaching area, metal forms holding, as if magnetically, the central space to its site. The small neighborhood units spin loose from the big circle in a free, almost lyrical fashion. "Sweet disorder," according to one nineteenth-century observer, reigned. But sweet disorder alternated in disconcerting unpredictability with the clarity of the individual neighborhood sections. Order and disorder dissolve one into another for the pedestrian in motion as cottages are seen from the front in a clear picket-fencelike row, or from sides or rear in a jumble. "The cottages lie in unintelligible radiation from some imaginary center or huddled together according to some undiscovered law of affinity." "The streets and avenues of the forest city do not intersect in hard right angles, but wind among each other in fanciful curves and mazy circles. This arrangement at first confuses the stranger and then charms him." It was "artistic." "Its lanes are as crooked as Quebec's, its parks as open as London's, though not as large, its plain and handsome cottages lovingly nestled together."[22] Wesleyan Grove became a special residential environment which, growing out of nature-immersion and spatial orientation of the early camp meeting, continued the sense of otherworldly dislocation necessary for religious experience, and thereby created a form that was almost the American romantic suburb as it would later develop.

THREE

THE COTTAGES

BETWEEN 1859 AND 1864, a new American building type, the campground cottage, was developed at Wesleyan Grove. The building form grew out of, in part, the cottage built by Perez Mason in 1859–1860, but is different from the latter in so many ways that it must be understood as a vernacular building type invented by local carpenters. There were about 40 cottages in place in 1864, 250 in 1869, and 500 in 1880, and they made of the complex tenting ground a permanent community that retains its memorable architectural form to this day.[1]

The Wesleyan Grove cottage is a two-story rectangular building, its short side facing front and its facade articulated according to a fixed, formal arrangement of considerable aesthetic impact. A wide double door centered on the first floor is reminiscent of both tent openings and church doors. To each side of the entry is a small narrow window. On the second level, under the gable, another double door opens onto a balcony that projects over the entrance. Two different patterns of jigsaw scrollwork, one usually small and tight, the other more flamboyant, hang from the balcony and from the eaves, the latter as vergeboard. The openings are usually in a Gothic Revival

style, with pointed or ogive arches topping the windows and doors, or in a Romanesque style, with round arches. This places most cottages stylistically among the romanticizing medieval modes of the early and mid-nineteenth century—modes which, especially in the Gothic version, were enriched by a variety of associations. Notions about feeling and revelation-based religious experience, as opposed to Enlightenment rationalism, about the feudal Middle Ages as a kinder and more charitable era than the competitive, industrial, capitalist nineteenth century, about medieval building forms being close to nature, in which God is immanent—all were blended into a loose but suggestive web in the minds of nineteenth-century Americans.[2] Although the churchly meaning of the Gothic is clear, that of the round-arched Romanesque style needs elucidation. It could share in the web of medievalizing ideals of the Gothic or it could have conveyed a sterner, Old Testament flavor in a Methodist context. There were more Romanesque Methodist church buildings in Providence, New Bedford, and Fall River by the 1870s, although some major ones were Gothic. An article on church architecture published in the Methodist press in 1855 recommended both Gothic and Romanesque designs without any discussion of differences in meaning.[3]

Although there are also cottages with squared openings, round and pointed openings predominate and define the character of the grounds. Window and door shapes are also emphasized by hood-moldings, following the ogive or rounded shapes, tacked onto the walls a few inches above the openings. All of the cottages are constructed of a single layer of random-width tongue-and-groove boarding laid vertically, leaving faint vertical striations, visible from both inside and out, as still another significant visual element. The shingling on the buildings, now so common, was added in later decades and is the major aesthetic distortion that the cottages have suffered over time. Shingles take away much from the Victorian tone of the buildings by turning them into something common to the twentieth century, and thus better understood by modern cottagers: the beach house. The now ubiquitous front porches are also additions from a slightly later period, the 1880s and after, but are less destructive of the visual quality of the buildings than the shingles. Porches, with roofs and turned posts, changed the spatial character of the grounds by adding a middle space, visually open but functionally private. Porches attached to unshingled cot-

---

(*Opposite top*) Round arch cottage. (Vineyard Vignettes)

(*Opposite bottom*) Round arch and pointed arch cottages. (Vineyard Vignettes)

Pointed arch cottage with porch added. (Ellen Weiss)

tages can reasonably be viewed as an improvement, a softening of the terser original cottage form.

The structure of the Wesleyan Grove cottage is as specific to the building type as the exterior forms. Each cottage is framed by six three-by-four posts, three on each long side, a system that reflects the Reverend Gorham's instructions for tent framing, only with heavier members and, of course, for a two-story rather than one-story form. The posts rest on a platform which is, in turn, raised about a foot above ground on cedar posts. The second-floor framing and roof plates are fixed to the six posts by notch and dowel joinery in most cottages, nails in a few others. The lumber is almost always long-leafed yellow pine, a wood with a resin that hardens over time, making the joinery rigid. There is some documentary evidence that cedar, spruce, and hickory were also used.[4]

After erection, the light six-posted cottage frame was enclosed by a skin of single lengths of random-width three-quarter inch tongue-and-grove boarding, the single layer creating the vertical striations so necessary to the visual character of the buildings, inside and out. Remarkably, there is no additional bracing in the interior of the buildings—neither studs at expected places, such as on either side of the doorway, nor diagonal struts. The rigidity of the frame may be due to the strength of the notch and dowel joinery, the fixing of the joinery from the hardening pine resin, or the tongue-and-groove boarding acting as a shear wall. There is evidence from some barely legible stereoptican prints that the light frame was braced during construction, before the board skin was put on, by long boards crossing the interior space on diagonals. Windows and doors were cut into the skin after the walls were laid up. Pieces of wall remaindered from cutting out windows and doors could be battened on one side and used as shutters during winter closure, protecting the glazing during the long part of the year that the cottages were unoccupied. Cottage roofs were con-

Cottage interior showing framing restricted to corner posts and vertical boarding. (Ellen Weiss)

structed of two-by-four rafters, without purlins or ridgepoles, covered with horizontally laid boards and then wood shingles. The lack of a ridgepole suggests that the structure is more than an enlarged tent frame, even though the six posts suggest the idea. This structural system was common in the eighteenth century in heavy construction, but much less well known in the nineteenth century. Almost all houses from the seventeenth through early nineteenth centuries in Rhode Island, a major source for Wesleyan Grove, have a heavy frame without studs or diagonal bracing. The framing is visible from the interior because the lathe and plaster are fixed directly to the vertical plank sheathing. Similar construction has also been found in the Piscataqua Valley of Maine and New Hampshire and on nearby Cape Cod and Plymouth. The rafter rather than purlin roof of the Wesleyan Grove cottage also suggests a derivation from Rhode Island, which retained plank-wall construction later than Cape Cod, rather than from the nearer place. Most eighteenth-century construction on the Vineyard is of the better known Massachusetts Bay type, with studs backing up the exterior wall surface, providing additional support between principal framing members, and separating the lathe and plaster interior finish from the exterior wall.[5]

One also finds an allusion to the tents that once stood on the site in the cottage interiors. About 15 feet behind the entrance, but not necessarily relating to the middle post, is a wooden partition with an arch dividing the rectangular ground floor into two rooms. The wide, prosceniumlike arches are edged in molding—the only architectural decoration inside the cottages. Period photographs show the arches covered with curtains, these often pushed aside to reveal beds or a dining table in the second room. The front room was furnished as a parlor. Tents often had the same spatial division: a front room visible to passersby and a more private one behind, the latter hidden by canvas tent flaps. Sometimes the arrangement seems to result from joining two tents together, end to end.

The back room of the cottages had one feature that was not a part of tent life. Narrow stairs to the second floor were built against a side wall and enclosed by the ubiquitous vertical boarding. Furniture for the upper rooms must have been hoisted over the balconies and in through the double doors upstairs. The modern kitchens and bathrooms now at the rear of the cottages were not structurally integrated with the main unit, although often at least as old. Some back sheds may antedate the cottages, having been built for cooking or for storing canvas and equipment for tents. Sometimes kitchens were added later, the early occupants taking their meals in boarding tents.

Formally and structurally, the Wesleyan Grove campground cottage was a remarkable invention. It was attractive, memorable, complex in its references—"house," "tent," and "church." It was inexpensive, quick to build, and consistent in form, for a unifying texture to the grounds, while containing within its fixed visual system possibilities for individual variation and display. The unarticulated rectangle of the long side mediated between the frankly shacky jumble of sheds at the rear and the highly formal facade. The long sides seldom have openings, only the occasional lancet window near the front or, under the eaves, a flap-regulated vent to air the upper floor. The proportions of the larger dimensions appear to be constant; that is, although cottage widths may vary from 11 to 16 feet, the vertical and longitudinal measurements change accordingly, suggesting a builders' rule-of-thumb for dimensioning. Much of the humor of the grounds comes from the varied bulk of otherwise identical units, the implication being that cottages grew to different sizes from some common seed. Windows and door dimensions do not vary with the size of the cottages, however, leading to both the consistency and the humor. Just as children have large heads in proportion to their height, the small cottages seem to gape even more openly because of the large entrances.

Another significant formal device of the Wesleyan Grove cottage is the 90° angle of the gable, with its concomitant 45° roof pitch. The gable triangle is one third the height of the cottage and roughly half of the width of the facade. That is, it takes three half-widths to climb the facade, two to eaves level and one to roof peak, implying a vertical tripartite division of the building. This relationship, never exact or explicit, is contradicted by another division of the facade, more clearly expressed, in two parts. The bottom of the balcony marks the halfway point between sill and roof peak. There is a tension between the two sets of implied divisions of two and three, which is resolved by the stability of the 90° gable at the top. The classic comfortableness of this resolution can be tested, for there are cottages at later campgrounds with more acute gables (Yarmouth) and more obtuse gables (Framingham) that have neither the serenity nor the memorability of the Wesleyan Grove cottage.

The feature of the cottages which contributes most directly to their popular appeal, lending a tone of sweetness to what would otherwise be a solemn scheme, is the jigsaw trim, especially the vergeboard. Wesleyan Grove's gingerbread is varied, robust, ebullient, yet delicate, with pendants of inverted flame motifs or chains of loops or circles and a variety of drops, all offering a rich visual display to the pedestrian explorer. Unfortunately, as with so much on the site, the documentation necessary to assign designs to individual craftsmen is missing. Fifty or 60 carpenters were working in the grove in the 1860s and 1870s, and although a few patterns have some stylistic affinity, suggesting a single hand, most do not.

The formal feature of the cottages hardest to define is their strange scale. Although they are "little houses," they are not miniatures of houses. Rather they are an independent type in which ideograms of "house" are combined with ones of "tent" and "church," all layered so tightly as to be inseparable. Part of the binding force which created a fourth typology, the campground cottage, out of the earlier three (house, tent, and church), is the dislocating shock of several juxtaposed scales, which, though contained within the classically proportioned facade, are so odd as to give the building a memorable, even hallucinatory edge. Multiplied by 500, the resulting sense of strangeness extended to the grounds as a whole. The wide double doors, with their tent and church allusions, are not only obviously and absurdly large for the building width, but also wide compared with normal house doors, dwarfing the inhabitants. The scale contradiction so apparent on the ground floor is even greater on the second. Because the first-floor doors are flattened, that is, ogive arch made Tudor and round arch lowered to a straight lintel, the doors are different in shape, if not style, from the small lancet windows to each side. On the second floor, however, the doors are not limited by the balcony and are topped by the same shape, round or pointed, as the ground-floor lancets. Thus they are perceived as huge windows rather than doors, as if for the appearance of a giant. There is a conflict of scale, and implied size of inhabitant, between every opening in the facade and the cottage itself. And because the openings are so large compared with the amount of wall, the scale shifts assert themselves as vast voids, holes of great expression, which, combined with the assertiveness of the vergeboard and the balcony, make something totally different and memorable.

The Wesleyan Grove campground cottage, as a new building type of remarkable singularity in appearance and structure, deserves an exacting history of its development. Because of inadequate documentation, the rest of this chapter will be concerned with not the history that is needed but the best that can be produced. It is probable that the cottage was invented on Martha's Vineyard by local carpenters, combining architectural elements available to them on the site or nearby at the beginning of the 1860s. Newspapers noted several cottages in place by 1861 and there were at least 40 in 1864. The camp meeting's own records do not distinguish between lots leased for tents and cottages before 1864, making it difficult to find the pre-1864 examples, the first of the type.[6] Even with occasional names of early cottage owners from newspapers, it is hard to trace the early examples because of changes in numbering and confusing arrangements of the tiny lots. Portability also makes traces unreliable. Cottages were frequently moved and another substituted with the only evidence in photographs, none in the records.

Late in the 1860s, when the cottages were a well-remarked phenomenon needing a history, several longtime residents wrote that the first wooden building was a one-story hut of about 10 by 12 feet, built in the mid-1850s for the Reverend Frederick Upham.[7] This would have been a cabin of the type deplored by the Reverend Gorham as a "shanty," unworthy of a camp meeting. Nevertheless, Upham's cottage "gave a new idea at which others soon caught, improving on this first domicile of the kind both as to size and expense, till finally the beautiful cottage erected by Perez Mason and William B. Lawton appeared."[8] There are other candidates for

Mason-Lawton cottage (Perez Mason, architect, *c.* 1859) and Fall River Row. (Dukes County Historical Society/Edith Blake)

the first two-story cottage with artistic ambition. According to the *New Bedford Evening Standard* for August 17, 1860, the splendor of the tents was being eclipsed by four frame buildings in a row belonging to parties from Fall River. These four other pioneers are not identifiable today, but the "paradisical little structure" for Mason and Lawton has survived intact because of the continuous commitment to Wesleyan Grove of many generations of the Lawton family. "Rarely have we seen so much of architectural skill and of artistic taste crowded in so small a space as is here presented before us. No one style of architecture fully expresses it, but the gothic predominates."[9]

The Mason-Lawton house is a double cottage designed for two families and actually occupied by both for a short time. It has many features of the mature campground cottage, but is a contribution to the form, not the form itself. With a long gable roof parallel to the front, a wood-decorated fake chimney on the roof ridge, secondary gables penetrating the roof, and vertical boarding, it is a cottage *orné* in the tradition of A. J. Downing's books of the 1840s. Downing, a horticulturist from the Hudson River valley with access to British art theory of the picturesque, wrote widely popular books advocating medievalizing or rustic styles of architecture. Country houses were to blend in with romantic, natural style of landscaping to create a domestic mood with more "feeling" than he believed could be found in the dominant classicizing styles of the period.[10]

The Downing-like Mason-Lawton house, unlike the Wesleyan Grove cottage, is not derived in major forms from the tent. It does not have the insistent verticality in massing, nor the roof ridge perpendicular to the facade. In plan it is a far looser design, with a long front section, narrow middle part, and wide rear area for the integral kitchen. The middle section is split longitudinally by a central wall separating two dining rooms, one for each family. The front section is a single long parlor, but is divided in half by a double proscenium arch on a line with the wall between the dining rooms. The proscenium is reminiscent of arches inside the campground cottage but is perpendicular to the facade, not parallel. Each half of the long front room has its own entrance, one for each family.

These dual entrances can be understood as introducing the Wesleyan Grove cottage motif: double doors with narrow lancet windows to each side, balconies above, and a second set of doors leading from the balcony into the bedrooms. The small second-story doors open into dormers inserted into the flank of the roof. The openings at the Mason-Lawton cottage, however, are neither Gothic nor Romanesque, nor as large or defined by robust hood-moldings as those of the Wesleyan Grove cottage. The tops of the windows and doors are two straight lines angled to a point, without curves. The nearest place for Vineyard builders to find the full-bodied Gothic Revival openings they used on their cottages would have been New Bedford, with its Gothic Revival monument, the William Rotch house (1845), designed by nationally prominent architect Alexander Jackson Davis.[10] The Rotch house offered, as well, the configuration of a frontal gable with decorative vergeboard containing a ogive-headed door opening onto a balcony which projected over a Tudor main entrance.

It is the structure of the Mason-Lawton cottage which is probably the clearest contribution to the campground cottage. The frame is slight and spare with doweled joints and tongue-and-groove vertical boarding, practically identical to what was used later. There are a few intermediate posts, such as those to each side of the entrances, and the sheathing boards are of regular width, unlike on campground cottages. It may be a bridge between the long-lived plank frame construction of Rhode Island and the Vineyard, for the cottage was brought in sections by sailboat from Warren, Rhode Island, during the winter of 1859–1860 for its architect-occupant, Perez Mason, designer of the association headquarters and the preachers' stand, and William B. Lawton. It shared a constellation of decorative motifs—an individualized form of vergeboard, triangular door and window heads, and latticework balconies—with two other cottages on the grounds. The Manchester-Brittain "Wee Hoos" was built on a tent-type plan for A. J. Manchester, a Providence school principal, and stands today. The Sarah Cook cottage, destroyed in 1970, was built for Mrs. Cook and her father, the venerable Methodist layman Hezekiah Anthony of Providence. Hebron Vincent, in a little-known pamphlet of 1872, called this one the first cottage of artistic ambition, meaning that it may antedate the Mason-Lawton house. It is best known to us in a picture taken by Aaron Siskind when he photographed the grounds in the early 1940s.[11]

Some cottages, then, built between 1860 and 1864, combined all of the architectural elements available to local builders to produce the classical campground cottage: the artistic ambition, entrance and balcony motif, and vertical boarding over light frame of the Mason-Lawton cottage; the two-room rectangular plan of the tents with three posts to a side and gaping entrance; and the full Gothic or Romanesque articulation of the openings. Newspapers do mention a few cottages before 1864, such as the second-to-largest building standing in 1862 which belonged to C. W. Field of Providence, or the "neat structure" of the Reverend M. Bemis of New Bedford, but these have not been located. "Pretty miniature cottages in lieu of tents" were still the rare exception in 1862.[12] There may have been 20 new ones in place by August 1863 and at least 40 in 1864. Several papers said that wooden cottages were now cheaper than tents because of the high price of cloth.[13]

By the middle of the decade, descriptions of cottages began to dominate the lengthening newspaper accounts of the grounds. The meeting's own *Camp Meeting Herald* estimated that 50 were built in 1866 and said that all were "exceedingly neat and tasteful, reflecting credit on the architects for their novel and ingenious designs in trying to adapt art to nature." The *Standard* said that there was now a total of 115, plus 20 other buildings, "kitchens and other shanties." Little cottages of various shapes, colors, and furnishings were located on parks, avenues, circles, and in "sly little nooks," displaying the different tastes of the occupants. "More original designs have been brought out this year than ever before, the older cottages having been mostly patterned after each other." This account is generous with notes on cooking and eating arrangements, describing board fences about back areas to secure some family privacy. T. O. Ruchman of New Bedford had a platform to the rear of his house with a cloth roof (no walls), plus a kitchen placed so as to leave a two-sided courtyard, making a pleasant place to spend a hot afternoon. Nearby, John Kendrick and C. H. Titus of Providence joined forces to build a wooden dining house in back of their adjacent cottages. An account of 1870 mentions a tent frame being used for an independent kitchen and dining room behind, but not connected to, a cottage.[14]

By 1866, cottage construction had become an obsession with speculative overtones. Tales were circulating about quick profits to be made by leasing or selling them, and some campers rented out their wooden homes, taking humbler tent quarters for the season. The following year the association ruled that residents could lease only one tent site, excepting adjoining lots for expanded cottages. The rule does not seem to have been enforced. Cottage development now doubled back on its own history. The Reverend Frederick Upham, the man whose board cabin was thought to have started it all, got a full-sized cottage and reduced his historic hut to rear-kitchen status. The presiding elder of the Providence Conference acquired a Gothic building, thought to be the best in the grounds. In 1866 one of the camp meeting's most distinguished residents, the Reverend Moses L. Scudder, author of several volumes on Methodism, acquired one by dubious means, if one man's testimony is to be believed. He persuaded a non-Methodist Hartford neighbor, David Clark, to build a cottage on the Scudder lot and then claimed it as his own. The year before, Clark had been a guest in Scudder's tent and was, according to his privately printed account, moved to tears by the solemn occasion of the meeting. "There must be real enjoyment with the Christian believer which the world knows nothing of."[15] It was also apparent to Clark that the meeting would become a great watering place, like Saratoga or Newport, but moral and religious in tone. Scudder introduced his guest to a builder and agreement was made to build a large cottage for $500. The first payment was due May first of the following spring, the second on July first, upon completion of the building. Scudder handled the contract, because Clark was off to another camp meeting. The following summer, Scudder had himself photographed in front of the new building, moved in, and put it about that his parishioners had given it to him. Later he had a basement dug as a way, according to Clark, of physically anchoring the building to its site so that the man who believed himself the rightful owner could not cart it away. Clark could do nothing, except to publish his complaint and hope that history would set the record straight.

By 1866, the year the Vineyard press was praising the Reverend Scudder's cottage, fascinated descriptions of the scene were appearing in newspapers from a wide geographic range. Although the Vineyard, New Bedford, and Providence papers are best for specific issues of numbers or carpenter attributions, the New York press was leading in evocative prose. Thus, one can read of the new cottages in *The New York Times:*

> The cottages are simply elegant. Great taste is displayed in the architecture of them. They are two stories in height, painted good bright colors, with door plate and door bells. The lower part is usually the reception room, while the upper floor is divided into sleeping apartments. They are furnished throughout very nicely, some of them even elegantly, and have all the necessaries and comforts of home life.

Or, on another occasion:

> The cottages are very pretty, with Gothic arches and balconies; they consist of only two small rooms and are open at each end constantly throughout the day, giving one a glimpse of all that transpires within. The curtain partition is drawn aside, revealing two tiny beds in the inner room, while the front is luxuriously furnished with pictures, sofa, carpet, easy chairs, settees. Canary birds sing outside the doors. Tree trunks are converted into moss-draped baskets of flowers. Ivy drapes the white tents with a beauty beyond architecture. Circles of shells ornament the roots of trees, and I saw one valuable Tritona ensconced in a low-spreading bough and filled with flowers.[16]

The cottages are "in most cases furnished in the best possible taste—in some cases even elegantly. Handsomely carpeted, with pictures, books, pianos, melodeons, shell ornaments and other devices for ensnaring the eyes, they are, as may be imagined, very beautiful. When you add to these features that they are tenanted in most cases by beautiful and charming young ladies who sit on the piazzas in light and airy attire, and chat and sing, or promenade in the avenues and parks, it needs no other description to show what an Eden I have got into."[17] Decor and personal possessions, icons of domesticity, filled out the architectural form to create an ambience.

Wesleyan Grove grew mightily through the 1860s and 1870s and increasing human density meant more cottages. Hebron Vincent estimated 4,000 people living on the ground in 1868, with 10,000 to 15,000 more there for short stays during the two-week period which included the actual revival. "Everyone has cottages on the brain." A newspaperman in 1869 said that there were 700 structures—250 cottages and 300 tents. Tents, still more numerous than cottages, were taken by some as a conscious holdout for old ways. The residents of Fourth Street Avenue were said to be the most pleasant people around because they lived in this "primitive camp meeting architecture." On one occasion the Fourth Street Avenue group held a religious dedication for a plank walk in front of their tents, a ceremony that was witnessed by a thousand.[18] The Fourth Street Avenue tents would remain intact through the era of stereoscopic photographs, a kind of pre-cottage historic zone. One tent was left in 1914, the only one on the grounds.

It was, however, the cottages that were subject to bemused descriptions. "All the houses are bright little painted boxes of fairy-looking cottages." "What a wonderful collection of tiny houses strewn about helter-skelter, like toys forgotten by children." "Swiss and gothic cottages, resembling large bird houses—bright, clean, and cheerful" graced the land. One was described as "a happy specimen of that difficult quality often longed for but seldom gained, and beautifully appropriate for the locality, fanciful neatness."[19]

The cottages looked set in the grove for the purpose of growing big and being transplanted at maturity to Newport or Cape May. They were mostly uniform as to shape—a gaping entrance, wide as a church door and Gothic in form, a railed balcony and a pointed window over the door. "Upon the stoop or piazza dwell the inhabitants. It is a figure of speech to say that they keep open house in the Vineyard." The door exhibited the interior in its entirety. The whole front opens like a doll's house. The use of space was as carefully studied as on board a yacht. "The single room is gayly decorated with cheap oleographs, flimsy *bric-a-brac,* gaudy worsted-work and many-colored flowers. A lace curtain drapes a second yawning door after the fashion of a set scene in a comedy. This curtain is adorned with cunning and elaborate devices, constructed out of the sad-colored leaves of last Autumn."[20] This peep inside is confirmed by photographs. Commercial stereoptican cards show furnishings arrayed before the proscenium arches on

Fourth Street Avenue tents, with consecrated plank walk. (Dukes County Historical Society/Edith Blake)

the wall opposite the front doors. The arches were covered by layers of both transparent and opaque curtains and valences, with leaves and flowers pinned to the fabrics, vines growing from pots over the whole, and framed mottos and pictures or shelves with lamps and bric-a-brac to each side of the arch. The elaborate arrangements were background to still lifes of chairs and tables, distributed symmetrically in front. The central parlor table was laden with shells, books, vases, a lamp, and potted greenery, sometimes as high again as the table itself, sometimes trailing foliage to the floor. One view shows such a set piece in a tent, rather than a cottage, the tent having a

Cottage interior. (Dukes County Historical Society/Edith Blake)

wooden divider wall (with a proscenium) to support all of the trappings. These constructions, visible from outside the cottages when the doors were open, seem to have a density of meaning, an iconic quality, which fixes cult objects, within the cottage-shrine, the beloved appurtenances of family and home.[21]

The richness of nineteenth-century prose evoking the cottage scene fails us utterly when it comes to architectural color. Accounts are full of references to "bright colors" or "many colors," but not to specific hues. The *Standard,* for example, reported "tasty new cottages painted in a variety of colors, all of them entirely different from any other in the ground," without being more specific. There is one reference to a cottage being painted pearl white inside and out. Another reports that some cottages were no color when fresh, but also mentions a group of new cottages going white. Some have suggested that the buff, ochre, and tan tones common in the mid-twentieth century may replicate older colors, traditional hues simply being maintained. One cottage was once painted several blues on the outside and plum, mauve, and pink inside, a style corresponding more to a late (1882) account of "fawn, plum and olive."[22] These are Queen Anne colors of the 1880s, not the pale earth tones sought by A. J. Downing for the rustic American scene 40 years before.

There are other unsolved historical problems posed by Wesleyan Grove's cottages in addition to their unknown inventors and original colors. We cannot, at this time, separate hands of individual artisans working on the site, even though their names are known. It may be that the cutting, milling, and marketing of the vergeboard and hood moldings, where one would search for signs of individual taste or craftsmanship, were such that any workman could buy any unit. If decorative work was interchangeable, then we probably will never be able to attribute individual cottage types to particular carpenters. It is probable that the standard Romanesque and Gothic modes were used interchangeably by a variety of craftsmen. Charles Worth of Edgartown built a large Gothic cottage for Senator William Sprague of Rhode Island in 1868–1869, and a Romanesque cottage nearby for Providence merchant N. B. Fenner in 1867.[23] But a round-arched cottage seemingly identical to the Fenner house was built in 1867 for N. B. Brock of New Bedford by two carpenters from that town, Solomon Chadwick and Henry M. Allen. Chadwick is also known to have built a Gothic cottage for himself.

Aside from the preponderance of Gothic and Romanesque cottages, there are some distinctive substyles that may someday have carpenter attributions. Most uncommon is a group of about seven with unusual vergeboard and, in some examples, a remarkable structural variation. The group, which has been dubbed "Flame" for the somewhat orientalizing decorative work applied to the building skin, lost its best example, the Stanton-Saunders cottage, to fire in 1973. In a piece of plank-wall-construction bravura, the front third of this cottage had no posts on the ground floor, the strength of the wall depending completely on curving in plan the tongue-in-groove planking from the side of the building to the front. Thus the single width of vertical boarding, curved, is the structure without benefit of framing.

There were other somewhat distinctive cottage types, although none were so flamboyant as the Flame group. At least three rows of identical units can be spotted in old photographs. One, joining the Mason-Lawton house, had flat, segmental arched windows with hood moldings ending in vertical labels on brackets, a characteristic domestic motif of the period. These could well be the "Fall River cottages" of 1860. A row associated with New Bedford residents once stood on the north side of Clinton Ave-

Cottages on Trinity Park. At left is a cottage of the Flame Group with one curved corner, without posts. (Dukes County Historical Society/Edith Blake)

nue. Another, on the south side of Clinton, was identified with Brooklyn and Great Neck, Long Island. These had brackets instead of vergeboard and a particularly robust round-arch hood mold with a chain motif detail. At one time, two of the "New York cottages" were connected by a tower with a mansard roof and its own arched entry. The walls between the cottages and the tower were removed, leaving a single long parlor served by three gaping entries. This conglomeration belonged to Joseph Spinney, a retired sea captain from Long Island, whose arrival one season was greeted with solemn mock ceremonies involving neighborhood ladies got up as allegoricized virtues. Only one New York cottage remains on Clinton, though another, or one like it, now stands on Ocean Park, in the later resort of Oak Bluffs, next to the campground. Adorned with a porch of particular intensity, it belonged to New Britain hardware magnate Philip Corbin (see page 83).

Joseph Spinney did not have the only towered campground home. The Crystal Palace, built by Providence coal merchant Henry C. Clark, later moved to Oak Bluffs, had a tower, as did the house built by Perez Mason for himself after he left the now-modest Mason-Lawton cottage. All three towers had high mansard roofs, dormers, and cresting, and all stood near the point where Clinton joined Broadway, the main entrance to the grounds. There were also, eventually, two three-story towerlike cottages

A "New York cottage." (Vineyard Vignettes)

Entry area in winter, 1870s. Crystal Palace (1869) is on the left; Perez Mason's second cottage on the right. (Stan Lair Collection)

that lacked the ambitious roofs and detailing of the Clark-Spinney-Mason enclave. The tower which once stood in Montgomery Square had a shop on the ground floor that housed the photographers Woodward and Son. The only tower-cottage that still stands is Tall Timbers, its name for the single lengths of board that rise through the entire three stories, without a break.

Although failing us with inventors, paint hues, and individual styles, nineteenth-century accounts of the growth of Wesleyan Grove do provide the names of craftsmen who worked there. Hebron Vincent reported 60 carpenters and painters at the opening of the August 1866 season, all "busy as bees, from early morn till dewy eve, erecting and beautifying these rural homes."[24] He named his Edgartown neighbors: Henry and Cornelius Ripley, Charles Worth, Freeman Pease, and Ellis Lewis. The Ripleys and Worth built in quantity over a long period of time. Charles Worth (1810–1884) was an islander who converted to Methodism at age 16 and was one of the lessees of the campground in 1846. Worth traveled to the California goldfields in 1849 and to the Pacific coast a second time 10 years later, returning to the island in 1861 just as the cottage boom was beginning. One of his children became the minister of the Matthewson Street Methodist Church in Providence, one a Methodist missionary in the Caroline Islands, and one a seafaring ne'er-do-well—whaleman, beachcomber, professional Christian—echoes of whose life, always far from home and full of "backslidden" states and new conversions, survives in letters to his parents.[25] At the end of his life, Charles Worth was living year-round in the campground, working as a carpenter and real estate broker out of his home on Commonwealth Avenue. The cottage does not survive.

The cottage construction of Henry Ripley, Jr. and his six building kinsmen—five Ripleys and brother-in-law Freeman Pease—was also often mentioned in island records. Henry and Cornelius, for example, had framed nine houses by the middle of March 1866. Ripley, Hobart, and Co. built 10

Tall Timbers. (Ellen Weiss)

by May 1868. Four members of "Henry Ripley's Edgartown gang" housed a Fall River client in 11 hours in 1866, the four being Calvin Tilton, James Pent, Alvin Stewart, and "Fred who talked." Henry Ripley also built the "cottage hotel," Dunbar's, in 1866 in Montgomery Square. This was a two-and-a-half-story, 21-room building in light plank frame construction, with especially lavish gingerbread trim descending 2 or 3 feet from the eaves. It was enlarged in 1870 to 60 rooms and, like the Beatrice House or the Central House, lasted until 1957, when it was demolished for parking. Together with the cottages, the old boarding tents were also becoming solidified into permanent structures. In 1865 Mr. Cody of Providence built a two-story structure with dining on the ground floor and lodging on the second, and Mr. Shore of Fall River built another. Boarding stations served 50 to 400 people in relays. Eventually there would be five hotels in the campground; one—the Wesley House—remains.[26]

Henry Ripley, Jr. (1820–1897), builder of Dunbar's, learned the housewright trade from Sirson P. Coffin, campground superintendent, housewright, and lumber dealer. A collection of Henry Ripley plans for a dozen farmhouses built from 1841 to 1850 is preserved at the Dukes County Historical Society. The houses are all of a standard island type, two-story end-gable three-bay Greek Revival buildings with the entrance left or right of two front windows, paneled corner pilasters, simple architrave and pilasters framing the door, and architraves on the long sides, under the eaves. Variations are in the plans and in the relation of separately roofed, low, rear kitchen wings (with brick ovens) to the main block. The consistency in proportions and massing of this house type, with its standard array of openings in the facade under a near 45° gable, the contrasting long plain sides, the rectangular plan, and the telescoping additions at the rear, is an argument for the Wesleyan Grove cottage as a vernacular development out of standard house form as well as out of tents. Ripley's design approach, if one dare interpret from this little collection, is of a competent and possibly inventive method of working within an established building tradition. His drafting style, ruled and filled in, is careful rather than artistic, but the houses have a balanced, classic attractiveness, not unlike the campground cottages.

Henry Ripley moved to New Bedford in 1867 and went into partnership in a lumberyard and planing mill. As member of an island firm, Ripley, Hobart, and Co., he kept up Vineyard work, doing so with distinction according to the *Gazette* in 1868: "Messrs. Ripley Hobart and Co., of this town, the justly celebrated cottage-builders, have been busily at work at their workshop near the North Wharf. Here may be found the trimmings for ornamenting the ten cottages which they have already contracted to build prior to the opening of the meeting in August. Some of the designs for trimmings are very tastefully executed, and reflect credit on the ingenuity of the firm."[27] The New Bedford operation might have produced island-style trim, although only a few examples of Wesleyan Grove hood molding and vergeboard can be found in that city today. Island gingerbread was probably winter work for jigsaw owners in the back sheds of Edgartown, at least until 1871. In that year the Fisher and Huxford planing mill was built next to the new Highland Wharf north of the campground. Fisher and Huxford, according to contemporary advertisements, imported not only the ubiquitous yellow pine, but also cedar, spruce, and hemlock, and produced tongue-and-groove planed boards, rough and planed joist, cottage frames, and shingles. They also had a jigsaw and a turning lathe. In 1880 Fisher and Huxford were joined by still another enterprise, the Cottage City Novelty Works, which also did complete cottage construction, including jigsaw trim.[28]

Five other Ripleys working in the campground seem to follow Henry in importance, although they too did a lot. Two of Henry's brothers, William (1814–1894) and Frederick (1824–1903), worked at Wesleyan Grove, as did William's son Joseph, later a successful contractor in Providence. Henry's other brothers were mariners, after their father, but one of them produced two carpenter sons, Cornelius (1820–1896) and Alonzo (1835–1912), both with many cottages to their credit. All of the building Ripleys were cottage owners as well, but no one style characterizes the houses that can be identified with the family.

Names of cottage carpenters from off-island also appear in the 1860s and 1870s. The campground was a vast arena of construction with, after 1879, building at the new, adjacent resort Oak Bluffs increasing the action. "These far-famed miniature cottages are springing up in every direction. In fact the number of buildings is limited only by the scarcity of workmen. The most indifferent mechanics, who can only occasionally hit a nail on the head, command $5 a day for their unskilled and inaccurate services."[29] Off-islanders mentioned in one year of the *Gazette* include Hathaway, Wilburn, and Bowman of New Bedford; Solomon Chadwick and Henry M. Allen of New Bedford; Chick and Sherman of Taunton; William Burt of Taunton; Mr. Hull of Charleston; and Green and Bates of Providence. Another year produced Mr. Marble of Fall River, who built the largest of the three mansard-roofed cottages in Cottage Park; Charles Talbot of Taunton, credited with a "new model" of cottage; Allen of Newport, builder of the two smaller "French roofed" cottages in Cottage Park; and Francis and Caleb Hammond of New Bedford. Not mentioned in the paper but fortunately remembered by descendants, is Robert A. Sherman of New Bedford. His

Stela Verda (Ellen Weiss)

own Gothic cottage, later named "Stela Verda," has a personal, even eccentric flavor. Because the doors have low-springing arches and consequently are narrow near the top, a full ogive arch could be fitted under the balcony, rather than the usual flattened Tudor arch. The loss of stability from the almost-square flattened entrance as a central void is enlightening, for Sherman's cottage has a surprised expression, a too-vertical emphasis, and is ultimately unsatisfying.

One designing and building islander does not appear at Wesleyan Grove, at least in this trade. Frederick Baylies, Jr. (1797–1884) was the only builder of an earlier period whose work merits the term "architecture". Baylies produced, in addition to many houses, impressive churches for all three active denominations, the Congregationalists, Baptists, and

Methodists, the last the "whaling church," today Edgartown's Greek Revival monument. He was one of the three original lessees of the campground and was still tenting in the 1860s. He refrained, it seems, from applying his fine designer's mind to the campground cottage.

What remains astonishing, ultimately, about the building record at Wesleyan Grove is not the numbers of carpenters working there and the wealth of individual variations which they brought to the cottage theme, but the degree to which they held to the type. The 45°-pitch of the gables, the symmetrical facades, vertical striations of the wood surfaces, the cantilevered balconies, the nature and placement of trim, and the overall scale and massing lend a consistency, a sense of unanimity of purpose or consensus to the enterprise. The 500 cottages of 1880 were more alike than different. All hands seem to accept that repetition and harmony were needed to maintain order in the burgeoning new grounds.

FOUR

WESLEYAN GROVE IN THE 1860S

By the end of the Civil War, Wesleyan Grove was changing from one kind of memorable community, a city of tents, to another, a city of cottages nestled under the trees. The change was startling, much remarked and described, and it earned a range of responses from observers reflecting larger attitudes toward growth, materialism, nature, and the designed environment.

Initially, at least, some of the older denizens of the camp meeting found a continuity of key elements, which left the sacred grove's traditions firmly in place. Thus, in 1864, in a group of 40 or more cottages, Hebron Vincent could still sound themes common a decade earlier and argue, as well, that the revival's religious power had actually increased with the passage of time. Dwelling for a few weeks beneath the shade of the grand old oaks in this place so hallowed by memories of early camp meetings, mingling with friends, walking and sea-bathing all promoted Christian health. It was better to come here than to spend more money, miss the religion, and mingle with baser folk at Saratoga or Niagara Falls. By 1867, the number of cottages multiplied fourfold, and Vincent's reports took on the dazed quality of those of the mainlanders who were seeing the grounds for the first time. "Is it a reality? Am I really in the old Wesleyan Grove or am I in some fairy-land?"[1] The oak forest which he had entered 32 years before was now populated by dense crowds. Instead of a few hundred humble seekers in rough tents, there were now many of fashion and wealth. Vincent found no conflict between the earlier and later eras, and his love of the meeting included all phases.

Other long-term participants were also amazed and welcomed change. All was growth, progress, and prosperity, and this meant not only worldly success, and power for the denomination, but also increased love of new generations for the faith of their fathers. The rudeness had passed away and pilgrimlike severity had mellowed into revered beauty. "We stood amazed and said 'It is beautiful; and it is splendid! It is truly a great aesthetic and religious institution.' The sight of the eyes affected our heart, and excited our religious emotions" at this Feast of Tabernacles.[2] This was 1868, when there were 200 cottages. Now it was an "aesthetic" as well as a religious institution—the architectural beauty mapping an experiential route from vision through feeling to faith.

But the new cottages also inspired debate over spiritual versus material values—for those who chose to worry about such matters. Vincent, typically, could accept material progress as inevitable and benign. Noting the unhappiness of tenters who were moved from long-held campsites to make way for cottages, he argued that, as in a growing city, wooden buildings, even old churches, must give place to bigger stone and brick works, so in this city in the woods "our house of cloth will be doomed to be superseded by one of wood, rising in architectural beauty."[3] Personal attachments to individual sites must give way. But stronger dissension also surfaced. In 1863, SAM, a correspondent of the New Bedford *Standard,* complained of the attention-getting richness of the family tents and cottages while the old society tents, where people professed to worship God, were without paint, without floors, with only benches and trunks for seats. "The walls cry 'O, my leanness, my leanness.'" SAM likened the elaborate cottages to a boa constrictor corrupting the spiritual tone and pricing out the poor. "When the purse rules, the common people mourn." This was, he thought, an era which carpets the parlor and kitchen and leaves the aisles and altar of the church bare, that "lives in a palace and worships God in a barn." His complaint was dual: Display makes the religious event too expensive for the poor, and rich people neglect their souls. But another *Standard* correspondent, Kirwin, disagreed. Kirwin thought SAM had a spleenish spirit and that if he would follow the owners of luxury tents to their mainland homes, he would find them lavishing attention on their churches there, too.

If architectural display in churches was all right in the city, why should it be a sign of degeneracy at Wesleyan Grove? Midcentury Methodists were often poor people who had gotten rich and this debate was common: "As the sect grows more prosperous it is getting ashamed of its ways and becoming less religious" versus "The camp meeting has caught the spirit of the age and with the church in general is marching forward in the path of progress."[4]

Others continued to object to the expanded and architecturally enriched scene from several points of view. Edgartown's *Vineyard Gazette* editor James M. Cooms, Jr. complained that the separation from nature that came with houses destroyed the idea of camping out. To be sure, he argued, one was still in the woods and could still hear the branches of the majestic oaks thrashed by the winds, but still it was not like lying upon the earth amid the fresh, abundant straw, with your boots or your coat for a pillow, lulled to sleep by the wind. This was the romance lost to those who were pent up in their snug cottages. Two years later Cooms again protested that the character of the canvas city in the woods was being lost to the wooden tenements, raising this time the religious value of the whole meeting. Returning home after a brief "Wesleyan campaign," he questioned sacrificing the comforts and conveniences of a well-ordered home for an irregular and tiresome life amid the haunts of poisonous mosquitoes and under the arbitrary rule of a select few. Instead of the earnest words and practical zeal that inspired works of grace and goodness, one now heard merry jests and rallying around the croquet ground, "a game, we are told, that tends to the softening of the brain."[5]

Even the meeting's own occasional paper, *The Camp Meeting Herald,* reported dissension. B. P., of the new Bedford County Street society tent, had a long memorv of early revivals where children spoke in a marvelous manner, an Indian boy of 12 holding a multitude in rapture for two hours. Of the old-style meetings, he remembered an artless simplicity and spiritual sublimity. "No night on earth seemed more lovely than was presented in the calm of summer midnight hour to one of these 'dwellers in tents,' who arising from his pallet of straw, would go forth to 'walk about Zion.' "[6] Now splendid cottages of architectural beauty, with the furniture of city life were supplanting the tents of Israel. He noted a post office, a telegraph, secular press, and even a broker's board recording the rise and fall of gold and oil prices. "All is bustle. . . . The carpenter still plys his hammer, a din of life and the eager chase for gain runs through the encampment." Yet B. P. also admired the cottages. He viewed them as neat and tasteful, a credit to the architects, and he was quick to point out that the Vineyard camp, unlike

Newport and Saratoga, did not fleece people in the name of fashion. Even this lover of simple ways could find some virtues in the new era.

Aside from the loyal Hebron Vincent, it was mostly those who did not know the meeting in its primitive phase who admired the expanded grove—its vastness, its strangeness, and its mixture of natural and material splendor. A visiting New Yorker felt a sense of enchantment grow with every turn about the place. Fairylike cottages succeeded picturesque tents, twisting and turning in every direction. Perfect order and quiet ruled. A grove of oaks shaded the seats through which sunlight flickered, making leafy outlines on the vacant benches, and the tones of the preachers' voices came softly in a continuous murmur. "The organ in front of the stand sent up a flood of melody and hundreds of throats sang 'I am happy, I am happy, I am on my way to Zion.' Happy? Well, they looked so at any rate—cool, comfortable, and 'happy.'"

Another New Yorker who had seen many camp meetings found himself unprepared for Wesleyan Grove. This one was a "figment of the brain, an effort of a distorted imagination." Personal experience, a mainstay of Methodist teaching, was necessary to understand its astonishing originality, its peculiarity, and its beauty. The phantasmagoric quality of the grounds was created by the multiple scales of the cottage facades and the abrupt relation between the cottage fronts and the public paths, distortions of normal house and community forms. Even George Francis Train, the notoriously eccentric promoter and author, found Wesleyan Grove bizarre and exotic. For him it was a "gigantic picnic, a grand fête cold-water . . . a monster tented tea party." Train boasted of world travels to such exotic places as a Chinese prince's hanging gardens and to tenting grounds in Turkey, China, India, Russia, Arabia, and Omaha, Nebraska. But none of these compared with the extraordinary scene at Martha's Vineyard. But then, after a thought about how much money he could make on the adjoining land, Train concluded that what was most remarkable was the modesty, purity, and "Arcadian piety" of this spiritual Newport. "I thank God there is some virtue left. The world is not all bad. Society may be organized hypocrisy, but not at Martha's Vineyard Camp Meeting."[7]

It was the combination of ostentatious materialism, eccentricity of form, and intense spirituality that impressed everyone. A New Bedford observer, writing in 1866, talked of the presumed trebling in financial worth of the grounds. Expansion had brought new wells, wharfs, roads, 50 more cottages, and new society tents, including one for a Newport group which had a wooden floor and sliding board partitions. Profiteering on cottage construction was widespread. Yet in the same paragraph he also claimed that the spiritual tone of the gathering was higher than ever and that the people were visibly filled with longing for spiritual renewal. It was the centenary of Methodist preaching in America and heightened religious endeavor was widespread. A few years later a correspondent would raise cottage materiality to the level of a perfectionist social vision. People should enjoy the world and have all the fine things they can get, because the day is coming when one man will convert a thousand at a time, sin will cease, and there will be no more war. Christians, in particular, should enjoy all the luxury they can (such as their beautiful cottages) so that the world can see that Christians do as well as sinners.[8]

It is unfortunate that no published sermons from Wesleyan Grove exist, for one would like to know all of the ways the startling physical environment might have been used by preachers to drive home religious truths. Suggestions of how tents and cottages were worked rhetorically do appear from other places. Father Edward T. Taylor, the "sailor preacher," used tents in a prayer delivered at Nahant, near Boston, as early as 1827. For him the canvas aglow at night showed life as a tabernacle, "through whose thin walls the lamp of a holy soul shines clearer and brighter" as the walls themselves grow thinner with age. Death was but the stepping forth from such a tent into "those glories which have no dimming veil between." To sanctified natures it was "only a step into the open air" from a tent luminous with the light that shines through transparent walls. Decades later Father Taylor, ready to welcome the new, saw campground cottages as emblems of religious progress. Modern campgrounds with beautiful cottages are part of the present wealthy state of the church and its greater means of doing good. Now souls could be saved by the thousands.[9]

Another example of sermonizing with campground fixtures comes from the Round Lake Camp Meeting, near Troy, New York. Here can be found a different lesson than that usually taken at Wesleyan Grove, even though the Vineyard meeting was the direct inspiration for Round Lake. At the New York meeting, cottages were viewed as fragile and temporary, within the overall natural scene, rather than as material assertions of Christian power. The real architecture, nature—the "Almighty's ample skies above" and the forest of stately trees—made man's little works in cottages incidental, just as the true Christian life overwhelmed human organizations, especially quarreling denominations. The Round Lake meeting was an attempt to reunite the separated northern, southern, and Canadian branches of Methodism. High over all the temporary differences within the church waved nature's fadeless green. "Christianity grows right up through all Methodisms so."[10]

Nature overrode man's works at Wesleyan Grove, as well, though much bemused or scolding attention was directed at the cottages. The community always remained hidden within the foliage of the gray-green scrub oaks, narrow tree trunks crowding close upon the dwellings. Although urban images were used to describe its density—city of cottages, celestial city, metropolis of tabernacles—the city was always a forest city, the canopy of overhead greenery intensifying its privacy. One explorer found himself entangled among the trees of this "oaken bower," worrying about how furniture could be gotten past the maze of tree trunks and into a cottage. Another found the cottages so sheltered and secluded that they only peeped out from the luxuriant foliage, increasing the fairyland atmosphere and furthering the distance from the prosaic world.[11]

The overriding sense of nature's dominion increased still again at night, despite the physical insertion of hundreds of aggressively detailed cottages. Admirers still wrote of the darkness's unearthly beauty, the spectral glow of lamps from inside the tents mesmerizing the seeker. As late as 1869, when there were 250 cottages to attract the debate of moralists, there were 300 tents, white by day but a haunting ethereal glow by night. One could still wonder at the "celestial city's pearly gates, whose translucence would manifest the beauty of the glorious light within. And is not Heaven begun below in most of these temporary habitations?" If one moved near the shore, the sounds of lapping water mingled with those of rustling leaves and the sweet, high voices of the distant singers to make a "phantasmagoria of light and beauty," just as in the primitive, pre-cottage era. Even in 1869 a wakeful seeker, suffering from awareness of his sins, could find in the glowing tents fresh hope that his struggle to break the chain of a selfish nature would end in victory.[12] The lamps of night still offered a route to redemption.

Still another aspect of the tented camp meeting which survived the escalation of wealth and materialism was its capacity to bond the participants into a group, like the tribe of ancient Israelites. Forest revivals, according to Gorham, had always aimed to strengthen relationships, and thus religion, among a dispersed people through joyous social interaction. The cottages seemed to have been designed for this communitarian purpose. The vertical striations of the boards and the overall vertical proportions made each cottage seem designed for close placement, one next to another, like pickets in a fence. The front platforms or piazzas and the visibility of the inside rooms established multiple plazas, like the apron and stage in a theater, for families to watch other families, make contact, enact domestic rituals, and breed a community of feeling. A social spirit was fostered and the reserve and isolation of "consequential" society worn away, leaving "fine feelings and emotions," together with religious principles. All who came there would become better just for being there. This is the kernel of modern perception of this unique environment. Wesleyan Grove sometimes appears to be the planner's answer to the 1960s communitarian complaint about the isolation of the nuclear family, the breakdown of society, and the retreat from the resulting anxiety into vicious competition and greed. "The grasping spirit of worldly gain does not intrude within the pleasant circle."[13] The layout of cottages about lanes and parks was in itself an antidote to the tensions of the age, tensions that were both the cause and result of materialism. Expensive cabins meant not only Christian power and worldly success for coreligionists, but also, in multiples, a warm and loving society.

* ** *

THE OPENNESS OF FAMILY LIFE in the community of feeling had another side that was more likely to catch journalistic attention—a lack of privacy. Travelers from cities, not understanding the communitarian aims of forest revivals, were highly amused at the display of family life. Beds and bedding, as well as living rooms, were visible from the paths, and much was made of how the art of doing up one's hair or strapping on a boot could best be studied here. Families at table might as well be in Madison Square for all to see. There was a calm indifference to public scrutiny that amounted to the sublime. Cottagers nurtured styles of self-presentation, judged by the

Cooking in the campground. (*Harper's Weekly,* 1868.)

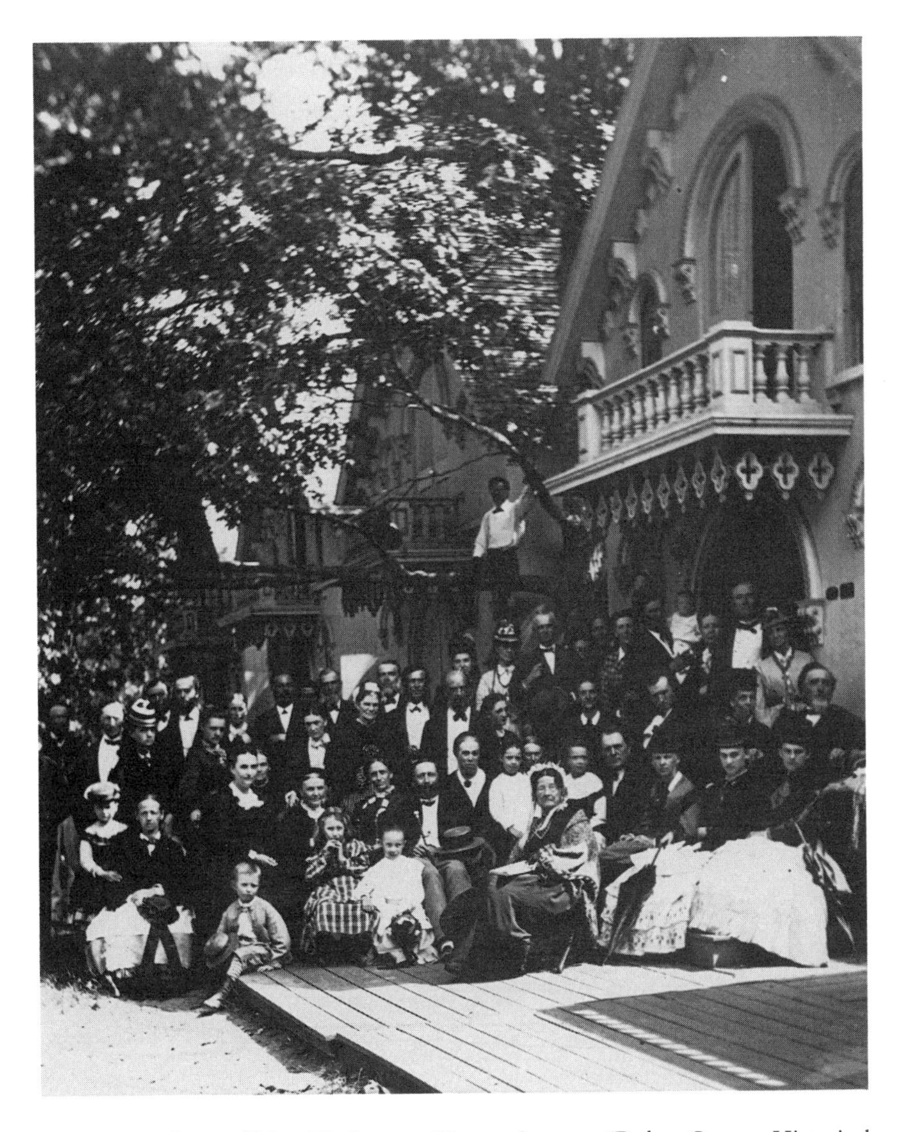

Cottagers in front of New York row, Clinton Avenue. (Dukes County Historical Society/Edith Blake)

artfulness with which they read or chatted in the evenings, so carefully lit by lamps. Families gathered as if posing for photographers—grouping was thoroughly understood, attitude everything.

Visitors were invited to stare into the cottages as they would into cages at a zoo while the inhabitants returned the stares with stoic indifference.

"Here we are! Come and look at us. We have nothing to conceal" seemed written over every portal. When a writer suggested that Barnum should hire the little city and show it in Central Park, one understands that he meant complete with inhabitants, an anthropological exhibit.[14]

The apparent openness of camp-meeting families with each other and consequently with the increasing numbers of day-trippers raised the problem of crime. Visitors from the 1860s on were amazed that there was so little theft of the possessions so artfully displayed on piazzas, trees, or within open cottage doors. Revivals had always been the targets of gangs. In the 1830s, a camp meeting at Franklin Grove, Illinois, was terrified to learn of the arrival of notorious highwaymen—the Black Legs—who, as it turned out, behaved perfectly, transformed by the spirit of the occasion. A similar exceptional mood prevailed at the Vineyard, or so it seemed. "Stuyvesant," in *The New York Times,* claimed that roughs, pickpockets, gamblers, and a few of the demimonde of New Bedford were held in order by the moral tone of the occasion rather than by officials. Wesleyan Grove was part fair, gypsy encampment, political convention, and barbeque but it would still be beneficial to anyone who came, even crooks. Arcadian innocence was realized; locks, bolts, and bars were unknown. Actually, by the middle of the decade, professional policemen from New Bedford and Pawtucket were helping the inexperienced island deputies with sophisticated off-island thieves who, for example, set a fire in the woods as a diversionary tactic while they robbed cottages. By 1867 there were 6 night- and 12 day-police, and a 12-foot square jail with a grated window was built as a warning. The completed building was soon in use and viewing the miscreant became an attraction. Even at camp meeting, Satan comes or sends a representative.[15]

Themes of Wesleyan Grove in the 1860s—the splendor and originality of the cottages, the dominion of nature, the openness of the cottages and the society to itself and to outsiders, the intense materialism and spirituality, and the lawful, moral atmosphere—came together in a single remarkable essay by James Jackson Jarves, a traveler, journalist, art collector, and critic.[16] For this visitor Wesleyan Grove was a successful though unknown Utopian colony. In 1870 Jarves had finished a frustrating and financially debilitating decade trying to lodge his collection of Italian primitive paintings in an American public gallery to help raise his country's aesthetic standards and artistic taste. He was a modest, retiring man who had suffered from the indifference and even hostility toward his collection of painting (which became popular later), and from financial loss of the investment. The year after his visit to Wesleyan Grove he had to sell his holdings at a sacrifice to Yale University, where they remain. Jarves's interpretation of the meeting as a Utopian antidote to the brutality of the age is, at least in part, a result of these frustrations. Wesleyan Grove was not, Jarves begins, the nomad's desert encampment that Methodists usually made. He had not led his readers into the wilderness to show them a prophet, but what he showed may be prophetic of radical changes to come in American civilization. Religious feeling pushes all people at some time to leave home and go into nature to worship. This urge turned Wesleyan Grove into a magical, wooded, "urban city." Hidden below the old oaks, was a grassy, mossy, plushlike loam, a soft, elastic carpet on which the extremes of age—first and second childhoods—could sport and tumble without harm. The trees cast down a cool, subdued light which alternated with mysterious shadows, a spiritual light like that of a Gothic cathedral. The branches of the trees were like the vaults of a medieval church. Avenues led from the entrance into the grove's heart, a temple of trees and leaves. Surrounding the temple, with its preaching stand and benches, were "tent-chapels." Jarves was comparing society tents to the side altars and chapels of Renaissance churches in which small services could take place in a more intimate setting. But croquet grounds were also close by, proving that strict asceticism had no place in this fresh phase of Methodism; the young here were taught to connect religious instruction with fun. Tiny children knew great freedom. Labeled with home tent addresses, they could wander the sacred city at will, safe as the lambs in Paradise, overseen by the general guardianship of a population founded like a family and cemented by religious faith. (Other sources confirm Jarves's observation: Tots were tagged with name, avenue, and tent number.)

While the inhabitants lived in their tiny cottages in a stage of perpetual picnic, the chiefs of the association managed all. Only those who agreed to the rules were allowed in, but the rules were few and simple. The Carpenter's Renaissance cottages sported flowers, seashells, wild vines, classical vases, and even sculptures. For this remote Eden had culture. Books from New York and London, lacquers from Japan, porcelain from China, and Parisian bibelots could all be seen just within the cottage doors. One need not fear theft for together with the prevailing goodwill, an invisible fraternal police watched over a neighbor's property as over his babes adrift in the woods.

> Here, even domestic life itself is as open as daylight. The reserve and exclusion which distinguish English homes do not obtain in this rustic life. Sauntering through the leafy lanes in close proximity to invitingly open doors and windows, one sees families at their meals, tempting larders in plain sight, and the processes of cooking, ironing, and other household duties, performed by the mothers or daughters themselves, with graceful unconsciousness or indifference to outside eyes. Occasionally, when cur-

> tains are not dropped, or sliding partitions closed, beds and even their inmates are disclosed. Everywhere ladies and children, in full or easy toilet, reading, writing, gossiping, amusing themselves at their discretion, unawed by spectators and as completely at home outside as inside their own doors.

The ease of the place was so palpable that Jarves claimed he could distinguish styles of rocking-chair usage: at Wesleyan Grove it was a *dolce far niente* tipping worthy of Italy in repose, as opposed to the usual Yankee nervous jerking motions. (Jarves had been living in Florence and had some special perceptions of the coastal New England of his boyhood.) He saw no newspapers in the grove. Their jarring politics and hair-raising or exhaustingly funny items were too much of a task for this sylvan paradise. This calm enjoyment, in contrast with ordinary Yankee restlessness in life, was itself a heaven on earth. Not even hordes of wonder-struck strangers, gaping at their most domestic privacies or vexing them with pertinent questions, perhaps impertinent too, caused the smallest ripple on their ocean of placidity. It was too deep to be moved by less than a hurricane. "For my own part, I felt as if dreaming. Had I got into an existence that had no reality outside of my own fancy?" He searched his memories of Europe, South America, and Polynesia for a similar "white day" in which man, nature, and art combined for perfection. The open cottages and their little piazzas reminded him of Pompeii, whereas the atmosphere, with its gravity and serenity, the sounds of the sea and the "primitive" habits and pleasures of the young, called back Polynesia, except that here the spirit was purer and less heathen. It was a true Christian fraternity, a hint of what modern society might be if it were not based on selfish and ambitious motives. "Here, on Martha's Vineyard, among the least imaginative and most hard-headed of races, was a social phenomenon which was worth investigating for its power of conferring substantial happiness." Thousands of average Americans of both sexes, in families or singly, had improvised a community with no police or law other than their free wills. Here they lived in social harmony, permitting the freest mingling of the sexes, ages, and classes while each individual kept his or her own character. These Americans camped in the wilderness, far from the great cities, in order to worship their God to the music of nature, and to give their wives and children pleasure after a healthful, sensible plan. These Americans found the spiritual repose for which everyone yearns, with enough intellectual and bodily activity to keep it interesting. Families did their own work, aided by restaurants and bakeries, so that a balance of labor and leisure was found for all without any of the difficulties that come with sharing homes and possessions.

The private and even commercial aspects of Wesleyan Grove were to Jarves an improvement on other intentional communities of theory and practice. "The question arises whether this sort of Camp Meeting Association might not be made the germ of others of an enlarged scope for city or country life," distributing the tasks of life and the talents of individuals in a more effective way than does present society, lifting its people onto a higher plane of intellectual and physical being. Perhaps only profound religious conviction had the power to bind such a community, but new communities could also be organized by cultural interests. Jarves seems to imagine residential arts institutions, with each member giving as he is best suited and receiving according to needs. "The bane of our civilization is its forced selfishness. Every man is obliged to become a rival of his neighbor in whatever department of life he chooses," succeeding more by his rival's mistakes than by his own efforts. "Our cities are crude experiments and perpetually recurring failures because they lack the central vital principle of universal good-will." Churches are selfish, too. Yet the Vineyard meeting was a successful experiment, so far as it went, in avoiding the haphazard, selfish bases of society in favor of the opposite principle of mutual advantage and goodwill.

No other observer made such a clearly Utopian interpretation of Wesleyan Grove. John Humphrey Noyes, historian of American socialism and founder of Oneida, once wrote in defense of camp meetings that the mixture of social, religious, and sexual excitements was good rather than bad, and that "divine" revivals required for their complement a "divine organization of society."[17] But only Jarves made the overt connection between the best of the permanent camp meetings and one of the age's great preoccupations, a perfectly structured civilization.

FIVE

OAK BLUFFS

FROM EARLIEST TIMES, the camp meeting's leadership showed an awareness of the ephemeral quality of its charge. Expressions of responsibility for the revival's fragile spiritual tone went hand in hand with admiration of its expansion, the mushrooming wooden cottages, and the meeting's increasing fame. By 1865, the year of the incorporation of the Martha's Vineyard Camp-Meeting Association and the land purchase, discussions were being held about buying the adjacent acreage to the south to protect the meeting's privacy. The campground occupied the wooded northwest slope of a low hill, the remaining area of the rise being open pasture. Ponds to the north and south of the hill, and 20-foot bluffs overlooking Nantucket Sound to the east, served as distant borders. The bluffs had long been important to camp-meeting life as a place of retreat from religious intensity and crowds. Inland from the cliffs about a mile was another acre or so of special meaning which was marked by a consecrated juniper tree. Traditionally, revival preachers had gone here for preparatory meditations, and it was also used for extra preaching space.[1]

The new association's notion of purchasing the entire hill, not just its shaded side, was postponed because the recent organization had exhausted everyone and because a conservative faction felt that land ownership was too worldly a preoccupation, smacking of speculation. The delay was a mistake, as a shocked membership learned during the revival of 1866. Six businessmen, four Vineyard whaling captains and two off-islanders, had bought the crest and the south face of the hill, including the bluffs and the consecrated tree. They owned 75 acres and were offering 1000 lots for sale. The religious community in the wilderness now had to share its necessary isolation with a commercial land development proclaimed as a summer resort.[2]

It would turn out that the new resort of Oak Bluffs, the camp meeting, and still another development, the Vineyard Highlands (on the hill north of the campground, developed by close associates of the camp meeting), would become a successful summer community with a religious flavor, carrying into the twentieth century values established by the camp meeting alone before 1866. In 1880 the three developments would separate from Edgartown, forming a new town, Cottage City, the name taken from Oak Bluffs's promotional tag, "The Cottage City of America." Renamed Oak Bluffs in 1907, the middle-class family resort, with a religious center and a uniquely festive spirit, would protect the old camp meeting with a shell of compatible residential development, allowing the campground to survive physically and institutionally, even without revival meetings. But this was not evident in the summer of 1866.

The meeting's first response to the surprise purchase was to consider moving to another site. Estimates were published that the campground facilities, including all of the cottages, could be transported to Chappaquidick for $7,000.[3] Chappaquidick, an island–peninsula south of Edgartown, offered a fresh wilderness, a convenient landing, Atlantic surf bathing, and a new stand of protective trees. The oaks which had shaded the revival since 1835 were dying, and this problem had raised the possibility of a move even before the developers had arrived. But the move was not made, in part, perhaps, because, as Hebron Vincent argued, the many advantages of the existing locale included its irreplaceable value as the spot where a great deal of good had already been done. Vincent never lost the sense of history of a place as part of its meaning and therefore of its power.

During the winter of 1866–1867, representatives of the camp meeting met with the developers and, presumably using the threat of a move as a bargaining chip, gained concessions that would assure some control. Deeds to property in Oak Bluffs would, and still do, contain provision that lots were for dwelling houses for families only, that the grantor had right of refusal on resale, that no liquor would be made or sold, and no gambling, manufacturing, or trading would take place. The state law prohibiting offensive trading near camp meetings would be enforced in Oak Bluffs during the period of the revival. These agreements were such that the *Gazette* could announce in April that the camp meeting would stay, but other problems had to be resolved, for feelings still ran high. The following summer the tent and cottage owners of Wesleyan Grove voted not to compromise or "make alliance" with the outsiders, and a year later even the non-Methodist campground residents voted for the construction of a high picket fence around the sacred precinct, to be closed at night and on Sundays.[4] The 7-foot fence was built, though the gates were seldom shut.

One of the first acts of the land company was to build a wharf into Nantucket Sound, a wharf that became popular with the revivalists because it was close to the campground. The old wharf in the Vineyard Haven harbor was more than a mile away. The campground responded by building a competing wharf at its new Vineyard Highlands development for an alternate tone of entry to the growing band and bunting festivity at Oak Bluffs. Arrival at the Highlands was a quiet, solemn affair, followed by a procession along a dike and bridge over the pond north of the campground. The road was named Kedron Avenue, after a stream outside Jerusalem, and entering the grounds from this direction would henceforth be known as "crossing over Jordan." Both wharves remained for years, forcing steamers to stop at each in turn.

But wharves as expressions of differences could become fields for resolutions. In 1870, just as the fledgling Oak Bluffs was beginning to boom, Erastus P. Carpenter, the leader of the six developers, acted in concert with the camp-meeting association by stopping steamers from landing passengers at the Oak Bluffs wharf on Sundays and closing restaurants on that day. This supported the Methodists on an issue of critical concern, the sanctity of the Sabbath. There were other signs of rapprochement. Hebron Vincent, ever the optimist where growth and progress were concerned, wrote that the campground would become the nucleus of a great summer resort, "a modern Bethesda attracting the best classes." Others were noting a new spirit of cooperation. Oak Bluffs's residents, for example, were praised for strolling on the bluffs instead of about the preaching area, leaving the religious seekers their privacy.[5] By 1870, it must have been evident that the developers were applying the style of the camp meeting, especially that of its physical planning, to their project. The plan and architecture of the new community, as well as its ceremonies and public life, were an extension of aspects of Wesleyan Grove. The development was going to learn from the revival.

The first way in which the developers showed that they intended to expand upon the campground, rather than conflict with it, was in the layout of the roads and lots. From at least as early as October 1866, the grounds were planned in a deliberately artistic manner, with a loose counterpoint of curves and countercurves, patterned as if the topography was rough, which it was not, or as if the style was derived from embroiderylike patterns endemic to the Victorian age. This is a method of subdivision which coalesced around the end of the century as the romantic suburb, but which, in 1866, had no clear precedent in this country other than the partially related Llewellyn Park, New Jersey, planned in the late 1850s, and the many popular rural cemeteries. Llewellyn Park, a private residential enclave with 6-acre lots unfenced to create the ambience of a vast country estate of movingly rugged beauty, was known in New England in the 1870s. A lengthy description in the Providence *Journal* is interesting for its interpretive slant: while the semiprivate development was modeled on an English country estate, the park was often open to the townspeople. Such housing hardly solved the problem of the slum-dwelling poor, but it was still an emblem of American opportunity for here no class system existed to keep people in their physical, as well as economic place.[6] Anyone could aspire to live there. There is no clear source for the Oak Bluffs plan in Europe, other than, possibly, Le Vesinet, near Paris, which was probably an inspiration for the beautiful Chicago suburb of Riverside, designed by Frederick Law Olmsted. Riverside was earlier in instigation but slightly later in design than its artistically more naive sibling on the Vineyard. Olmsted and Oak Bluffs's designer, Robert Morris Copeland, knew each other and must have shared ideas directly or as members of the larger culture of those who thought about habitation and landscape. Coincidences abound: Both the Chicago suburb and the Massachusetts resort had a business block built in 1870 called the "Arcade." Both had nearly contemporary resort hotels. Both must have grown out of general discussions about inward-turning neighborhoods and curving streets for moods of tranquility and ease. But Oak Bluffs is probably an independent invention, borrowing its intricate plan of curving roads from the rural cemeteries that Copeland was designing in the Boston area in order to be a sympathetic neighbor to the woodsy, mazy campground. Oak Bluffs is not derived from English Regency-styled resorts, as John Archer has argued for the American romantic suburb, for the English examples are more classicized, with clear geometric layouts which preclude the romantic complexity and wilderness allusions for which Americans yearn.[7] Americans want a more intense relation with a more primeval nature than their English cousins, as can be seen in the continuity from wilderness camp meetings to modern tree-shaded and bush-tangled suburbs of single-family dwellings.

Rural cemeteries, with trees shadowing little lots strung along curving roads, often in leaflike patterns, were popular in many American cities but especially in the Boston area, largely because of Mount Auburn in Cambridge—a cemetery, arboretum, park, and tourist attraction. By the 1850s lovely, sinuous, planted cemeteries were being laid out as institutions of civic and aesthetic progress all over eastern Massachusetts. One was Rock Hill cemetery in Foxboro, home of Erastus P. Carpenter, leader of the Oak Bluffs's developers. Carpenter, president of the corporation that hired cartographer H. F. Walling to design Rock Hill in 1851, was a man of many business and civic accomplishments. He had already introduced steam power into his family's straw hat factories and had welded these into the Union Straw Company, which employed 6,000 workers. His civic activity included a "New Jerusalem" of housing for these workers and restoration of the town common, rounding out a legacy of social and environmental concern.[8]

Carpenter then, must be the man who convinced Oak Bluffs's developers to find a landscape gardener experienced in rural cemetery layout to plan the lots and streets of the new village. Robert Morris Copeland had already distinguished himself by writing with H. W. S. Cleveland an early plea for a profession of landscape architecture, *A Few Words on the Central Park* (1856). The Copeland-Cleveland partnership of the 1850s produced rural cemeteries for Gloucester, Waltham, and Concord. Copeland, now on his own because Cleveland had moved to Chicago, had published a horticultural manual, *Country Life* (1859), in an effort to bail himself out of a failed farming venture. Oak Bluffs was Copeland's first subdivision design, antedating both his unexecuted plan for Katama, on the Vineyard south shore, and his well-known Ridley Park, near Philadelphia.[9]

Copeland's first plan for Oak Bluffs, inscribed October 1866, is an intricate scheme, loose in its configurations of curving streets interlaced with variously shaped small parks. This design was not widely distributed, only one copy being known in the twentieth century. That one was, fortunately, reprinted by Henry Beetle Hough in 1936, for it has since been lost. Oak Bluffs was built to a second design made in the early months of 1867 and redrawn by Copeland on the same stone as the 1866 scheme, the earlier inscribed date being left unchanged. This design is well known today from republication by John W. Reps, and was widely distributed in the nineteenth century. Copies were given to anyone stopping at the land company offices. Copies were framed and displayed in mainland hotels, on boats, and in railway stations. Small versions were used as a logo on com-

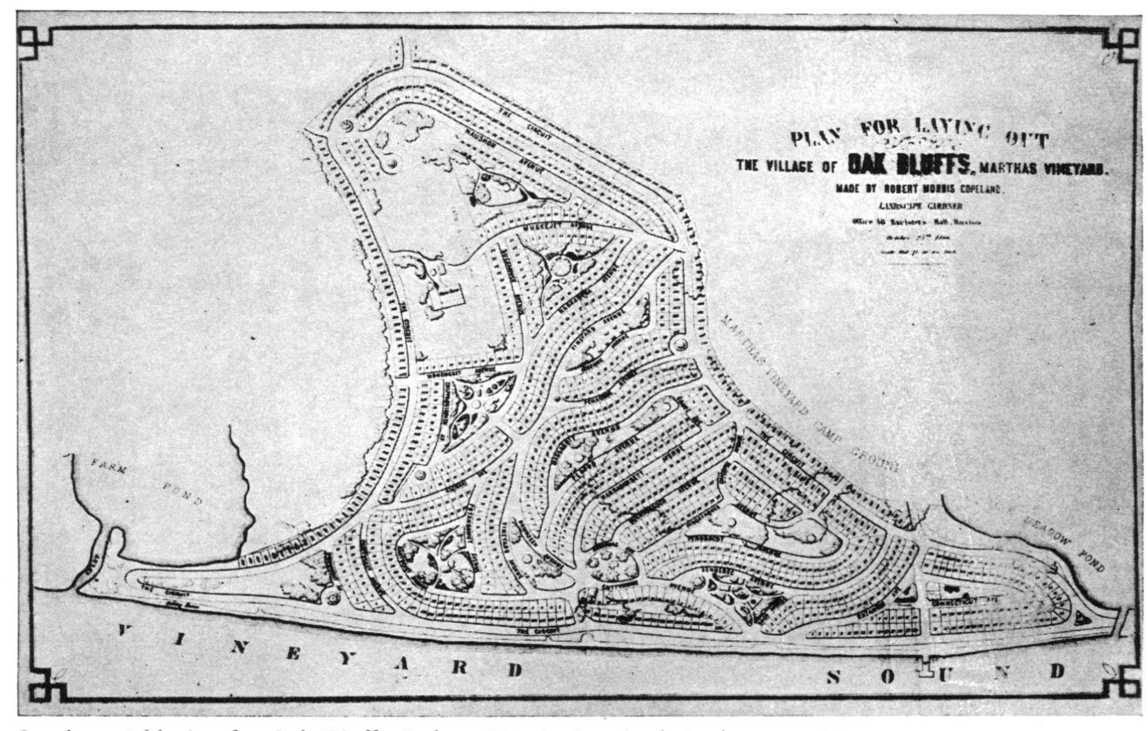

October 1866 plan for Oak Bluffs. Robert Morris Copeland, landscape architect. (Henry Beetle Hough, *Martha's Vineyard, Summer Resort,* 1936)

pany letterhead. This second plan incorporated several features of the first but also set out a dominating new motif. It sacrificed hundreds of salable lots in the area near the bluffs to make a 7-acre park, rendering the community open to space, sky, and the sea, perhaps at considerable loss in intimate neighborhoods about tiny parks, on the campground model. The new Ocean Park was the developers's idea, not the landscape architect's. Twenty years later, while giving testimony in a court case over park ownership, they provided their reasons. The land in that area dipped in the center, creating drainage problems for cottages. "Breathing space" and firebreaks were wanted. A large park would help avoid conflict with the camp meeting and would give the development "magnitude." And it would attract city dwellers and a better class of resident to its edges.

The last notion justified itself almost immediately. A group of Connecticut businessmen bought several adjacent lots at the head of the Ocean Park. Their lots were adjacent in the sense that they all lived near each other, a row of friends and associates, and in the sense that each purchaser bought several lots so that they could build large cottages. Their families became central to the life of the new community, and were on hand two decades later to testify for public ownership of the streets and parks, the issue to be settled. Their commitment to their summer homes, fleshed out by their social and material resources, set a direction for the resort and they seemed one with the town's history and with the park their houses overlooked: generous, gregarious, celebratory, self-assured, and clearly contained at the edges. Today the land sides of Ocean Park are still forcefully defined by the row of highly individualized, but closely spaced, cottages. As seen from the water and wharf, the houses of Oak Bluffs appear almost as a continuous form around the park. In the nineteenth century they were even more assertive, bristling with towers, cupolas, crests, finials, and pinnacles, the collective silhouette like a magical city or palace of the imagination, recalling Chambord or the skyline of Oxford.

The first plan of Oak Bluffs, that by Gopeland without the developers' big park, was for a roughly triangular mass with gently curving sides, a shape reminiscent of Martha's Vineyard itself. The longest side of the triangle, at the base in the plan but on the northeast on a map, corresponds with the line of the bluffs and the sea. One short side of the triangle, to the right on the plan and to the northwest on a map, is the boundary shared with Wesleyan Grove. The entire 75-acre area is encircled by a continuous road, The Circuit. Within the loop are 12 small parks of about an acre apiece, each a different nongeometric shape. Gently curving streets wander around and between the parks, creating an informal ambience for the leisurely stroll, the casual encounter of neighbors, and the shared savoring of the

(*Top*) Cottages on Ocean Park. "New York cottage" with porch added for Philip Corbin. (Dukes County Historical Society/Edith Blake)

(*Bottom*) Oak Bluffs and Ocean Park, *c.* 1890. (Dukes County Historical Society/Edith Blake)

languors of summer escape. The plan has none of the coherence and monumentality of Olmsted's Riverside, but it does have a deliberate sweetness and a mood of relaxation and ease. It proposed a way in which "natural beauty abetted by intentional charm" could shape the feeling of a residential neighborhood.[10]

Two elements from the first "October 1866" plan, that antedating the developers' intervention, actually were executed and survive today. The Circuit, the continuous loop road encircling the village, is there, if only in unpaved paths in the southernmost sections. That part of The Circuit which borders the campground soon developed into a commercial street with shops, restaurants, and large hotels supplanting a first generation of cottages. This is the present-day Circuit Avenue, the spine of the town. The section of The Circuit bordering the bluffs and the sea later became connected to roads to Vineyard Haven and Edgartown, losing its identity as a piece of the loop. It retains one, however, as a mediator of some ceremonial scale between the town and the sea.

The second surviving element from the 1866 plan is Hartford Park, named for the home city of many of its earliest residents. This long, sinuous shape of about an acre preserved Wesleyan Grove's consecrated tree and the adjacent preaching space, a gesture of sensitivity to the acquired meaning of the land and the feelings of the campground residents. Prayer meetings continued to be held here for four or five years, and it remained a closed shady retreat, used for decades for neighborhood entertainments from tea to concerts to croquet. Its form, narrow but about two blocks long, curving on one side only, may be the best remaining example of Copeland's style of this time.

The most distinctive feature of Oak Bluffs today is the developers' Ocean Park, a 7-acre, slightly concave space visible from the deck of a steamer approaching the wharf. The park is semicircular in plan with a slight triangulation at the apex and is somewhat barren, almost treeless. It is a surprisingly monumental space, partly because of its emptiness, partly because of the clear definition of its edges by the closely set cottages. Points within the expanse of sky-revealing land are fixed by ornamental ponds, small gardens and, from the 1880s, a bandstand. The roads shown on the plan as crossing the park in sweeping arcs were removed by 1871 and a rail fence, said to be modeled on the one bordering Lexington Common and designed to exclude horses but not people, was placed to make "the chil-

---

(*Opposite*) Final plan for Oak Bluffs, 1871, incorporating the 1867 plan. Robert Morris Copeland. (Suffolk County Land Court)

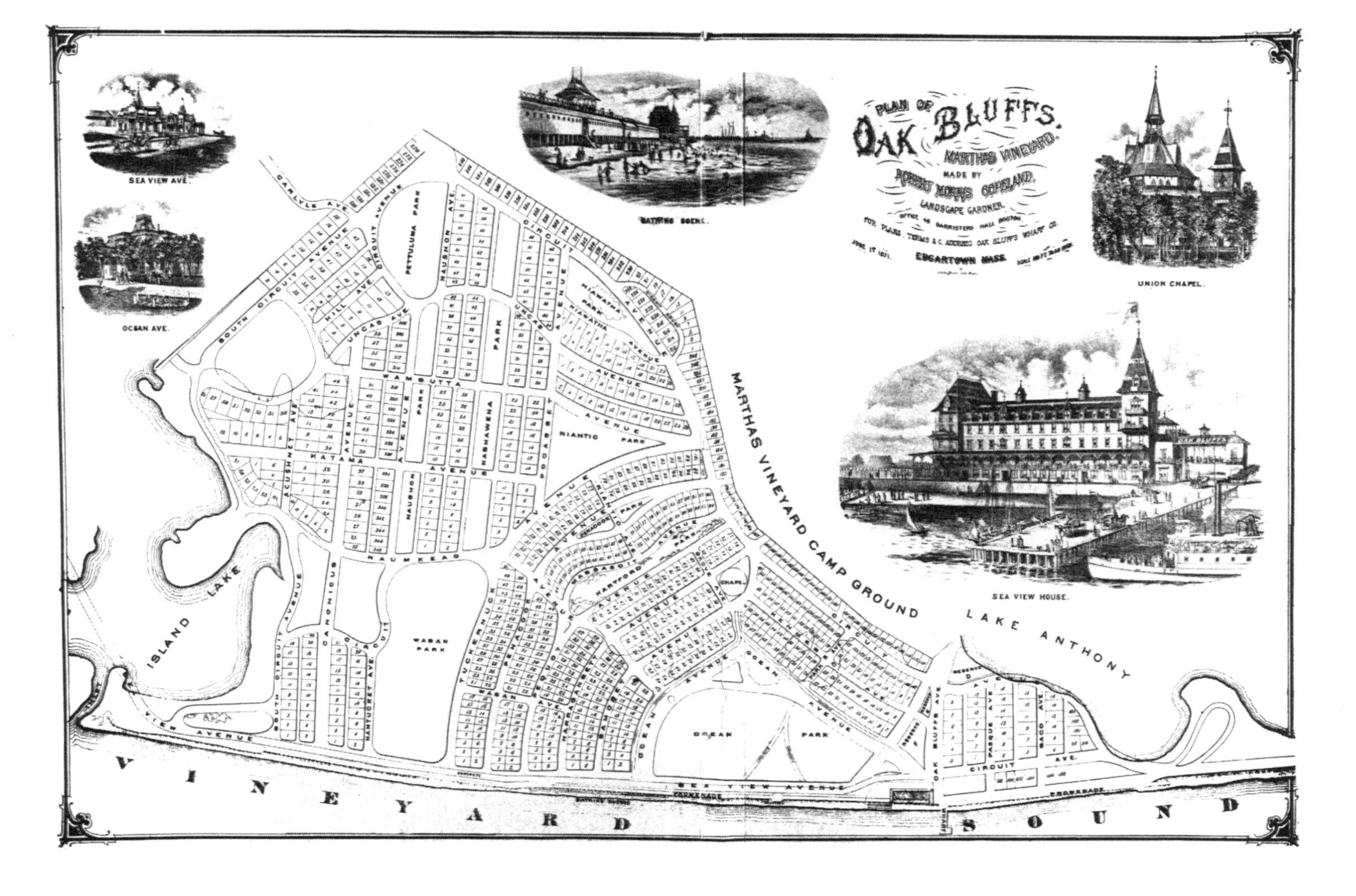

PLAN OF OAK BLUFFS, MARTHAS VINEYARD.
MADE BY ROBERT MORRIS COPELAND, LANDSCAPE GARDENER.
OFFICE 40 BARRISTERS HALL BOSTON
FOR PLANS, TERMS & C. ADDRESS OAK BLUFFS WHARF CO.
EDGARTOWN MASS.
JUNE 1ST 1871.
SEA VIEW AVE.
OCEAN AVE.
BATHING SCENE.
UNION CHAPEL.
SEA VIEW HOUSE.
MARTHAS VINEYARD CAMP GROUND
LAKE ANTHONY
ISLAND LAKE
VINEYARD SOUND
SEA VIEW AVENUE
PROMENADE
WHARF
OCEAN PARK
OCEAN AVENUE
WABAN PARK
NIANTIC PARK
PETTULUMA PARK
HIAWATHA PARK
CHAPEL
CIRCUIT AVENUE
SOUTH CIRCUIT AVENUE
CARLYLE AVE
NAUSHON AVE
UNCAS AVE
HILL AVE
WAMSUTTA AVENUE
KATAMA AVENUE
ACUSHNET AVE
NASHAWENA PARK
NAUMKEAG
CANONICUS
NANTUCKET AVE
TUCKERNUCK
POCASSET
HARTFORD AVENUE
OAK BLUFFS AVE
PARQUE AVE
SACO AVE

(*Top*) Looking north along plank walk toward S. F. Pratt mansard-roofed cottages and Sea View hotel. (Dukes County Historical Society/Edith Blake)

(*Bottom*) "Pagoda" refreshment pavilion. S. F. Pratt, architect, *c.* 1871. (Dukes County Historical Society)

dren's paradise" safe. The long diameter of the park, aligned with the bluffs and the sea, reads as a linear platform for the study of horizons and any thoughts that might attach to them. Early construction by the Oak Bluffs company included a plank walk the length of the bluffs, with a shaded "bath arbor" on the bluffs for watching the swimmers below, and a glass-walled octagonal refreshment pavilion. These structures, together with vast Sea View Hotel of 1872 at the north end of the bluffs, were carefully positioned to protect the water view as seen from the houses, directing the attention of the community to the horizon. Nantucket Sound was part of the main waterway from northern to southern New England and New York and it was crowded with sails. Parading the bluffs and watching the aquatic display became an occupation for vacationers at the resort, as it had been for revivalists before them.

The insertion of Ocean Park into Copeland's design carried with it the problem of transition from the new semicircular space to the older parts of the plan, such as Hartford Park. Either Copeland or the directors solved it by extending some moderately long streets from the 1866 design to make a concentric pattern of great curving arcs of roads. Samoset, Narragansett, Pequot, Pennacook, and Tuckernuck avenues are somewhat relentless in length and in fixity of curve, but, like Ocean Park, they provide a kind of "magnitude" which, though probably cruder than anything Copeland first intended, is also impressive. Similar long, even-curved arcs bordered by houses set back uniformly on small lots appear to be a conscious act of monumentality at Riverside. The effect is not quite so sure or intentional at Oak Bluffs, but it is there. Such regular placement of cottages on evenly spaced lots suggests an image of collective wealth, values, and social and emotional habits. Yet the individuality of each cottage also emerges from this picture as, perhaps, a too-easy metaphor for the Americanness of this middle class at play. One is reminded of the powerful crescents of English resorts and the relentless similarity of each house facade to the other in those places. Individuality as an ideal would seem to be getting some expression in the modest American version.

The 1867 Copeland plan for Oak Bluffs was modified again in 1870 and 1871 to accommodate two later land purchases totaling another 45 acres. Copeland's designs for the extensions show a different style than that of the 1866 and 1867 plans. Many of the new streets run straight, in elongated blocks with a green median park. The new formalism might reflect an admiration of Back Bay, with Commonwealth Avenue, in Boston, just as the 1866 plan might conceivably reflect an admiration for the confusions and psychic dislocations of old downtown Boston, to which at least one guidebook compared it. The elongated grid of the new section is contained within a curving leaf or fish form poised on a sloping hill looking down on a second large park (this too a last-minute developers' intervention) and the sea. The leaflike edge of an area of long straight blocks could have been learned from the now published Riverside plan, the line of the railroad in the Olmsted design playing the same formal role as the green median strip at Oak Bluffs.

It is probable that Oak Bluffs was laid out in this inventive curvilinear mode to make it a physical continuation of Wesleyan Grove, to persuade the revival to stay and to borrow the proven charm of the grove for the resort's own benefit. The feeling of Oak Bluffs is, however, somewhat different, its scale more like a normal residential neighborhood than is its otherworldly neighbor's. Streets are streets at Oak Bluffs, not paths. Lots are larger and building setback requirements created a normal relationship between house, expressed private land, and public road. Lots in the 1866–1867 section of the resort are 35 by 65 feet. Those in the 1870–1871 section are 50 by 75 feet. Campground lots were only about 14 feet wide in the older areas, and up to 20 feet wide in the newer ones. (Depth of lots in the campground varied according to location but were usually at least one and a half times the lot width.) Oak Bluffs deed restrictions required building setbacks of 10 feet from the front and five from each side. (There were no setback requirements from the rear lot line.)

The Oak Bluffs plan as executed is attractive, especially with the somewhat barren quality of the land from the paucity of trees, but it is not great design. The joinery between the different sections of the community is not smooth, and the loss in orientation which the traveler sometimes experiences often seems more accidental than intended. Copeland's 1872 plan for Katama, a resort planned for the southern shore of Edgartown but never built and another E. P. Carpenter enterprise, is more coherent and lyrical, with its long swinging arcs of roads and the counterpoint rhythm of short streets and slivers of parks fanning out toward the sea. Oak Bluffs seems naive and stilted when compared with either Katama or the magnificent Riverside, but it has the saving grace of any early art form, the charm and sincerity of a style on its way to being.

Even with the imaginative physical planning of this early romantic subdivision, success was hardly guaranteed. Lot sales went very slowly during the first two years. There were five cottages in place by August 1867, and 12 cottages plus 30 tents the following year. The company had built an elegant gatehouse at the head of the wharf, partly to hold back the crowds of revivalists who wandered down to the new wharf to watch the boats come in. The gateway, expanded in 1870 with a second-level balcony connecting a press room with a company office, (see page 114) was attached to a large company service building, a two-story structure with "Oak Bluffs" in big letters facing the sea. This first company building served for storage, lodging, and as a restaurant until it was removed in 1870 to make way for the Sea View hotel. Its image was revived only eight years later in a colored lithograph memento of the then-successful resort's beginning. The airy fantasy of the expanded gatehouse was the work of S. F. Pratt, the Boston inventor-architect of talent and verve who would be responsible for most of the land company's structures, and his partner John Stevens. Stevens designed for the company's directors several early cottages which had mansard roofs, towers, and a front wall which slid open to reveal the interior.[11]

It was not until 1869 that anyone could be assured of Oak Bluffs's success. Sixty cottages were constructed that year and projections about outstripping the campground began to be believed.[12] Most of the cottages in the resort were of the Wesleyan Grove type, reiterated by the Ripley gang, Charles Worth, and others from the campground. This cottage, especially its Gothic variant, was the basic building block of the resort, even when so overlaid with towers, porches, bay windows, dormers, and creative roof lines as to be almost unrecognizable. It is only in the few which do not have central entrances and in the group which can be assigned to S. F. Pratt, that one sees any significant change from the campground's building style. And even in these, the construction is in the usual light frame with tongue-and-groove vertical boarding, the plank frame system of Wesleyan Grove. The success of Oak Bluffs as a good neighbor and convincing formal continuation of the campground is due to its architecture as well as its plan.

By 1870, the developers must have been convinced of their success, because they invested in a two-year building program of large and small but always aesthetically ambitious structures. First came the Arcade, a combination of offices, shops, and a monumental gateway into the campground's commerical zone, Montgomery Square. The Arcade, with decorative Swiss-style exterior woodwork, a three-story step-back profile, and kicked eaves perking up the roofline, was placed midway on that section of The Circuit destined to become Circuit Avenue. The two-story "Oak Bluffs" building by the wharf was removed, eliminating the plain-style introduction to the resort and clearing space for a grand hotel. All of the company's buildings henceforth would show a new level of architectural ambition, from the airy fantasy of the wharf gateway, to the Arcade, the bath arbor, the Pagoda (the octagonal refreshment stand), the chapel, and the Sea View hotel. These and some 12 to 18 cottages, including one for the governor of Massachusetts, were designed by Pratt.[13]

Samuel Freeman Pratt (1824–1920) blossomed as an artistically mature designer in middle age after an earlier career, still largely unmapped, which

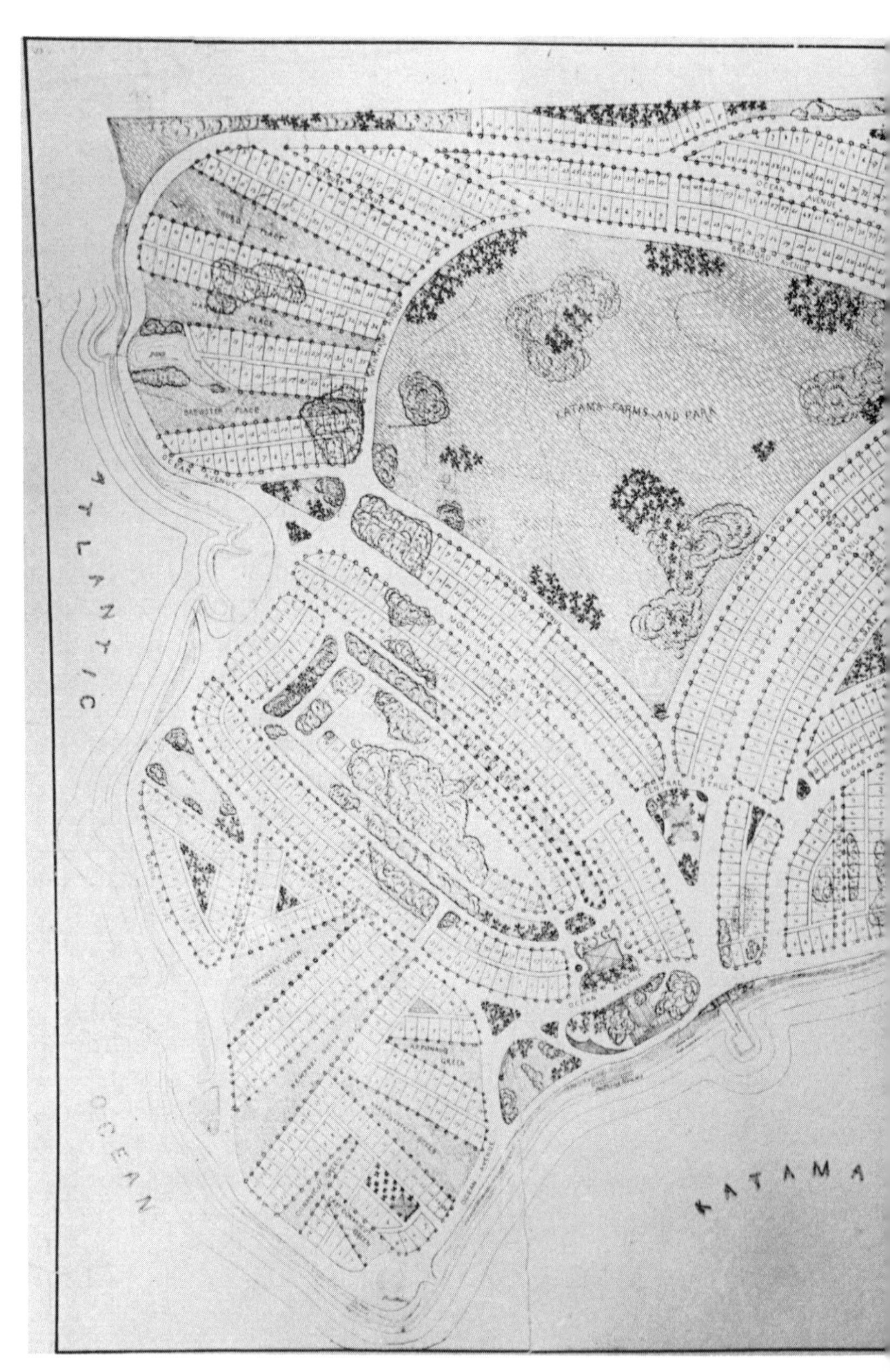

Unexecuted plan for Katama. Robert Morris Copeland, landscape architect, 1872. (Morgan Clark)

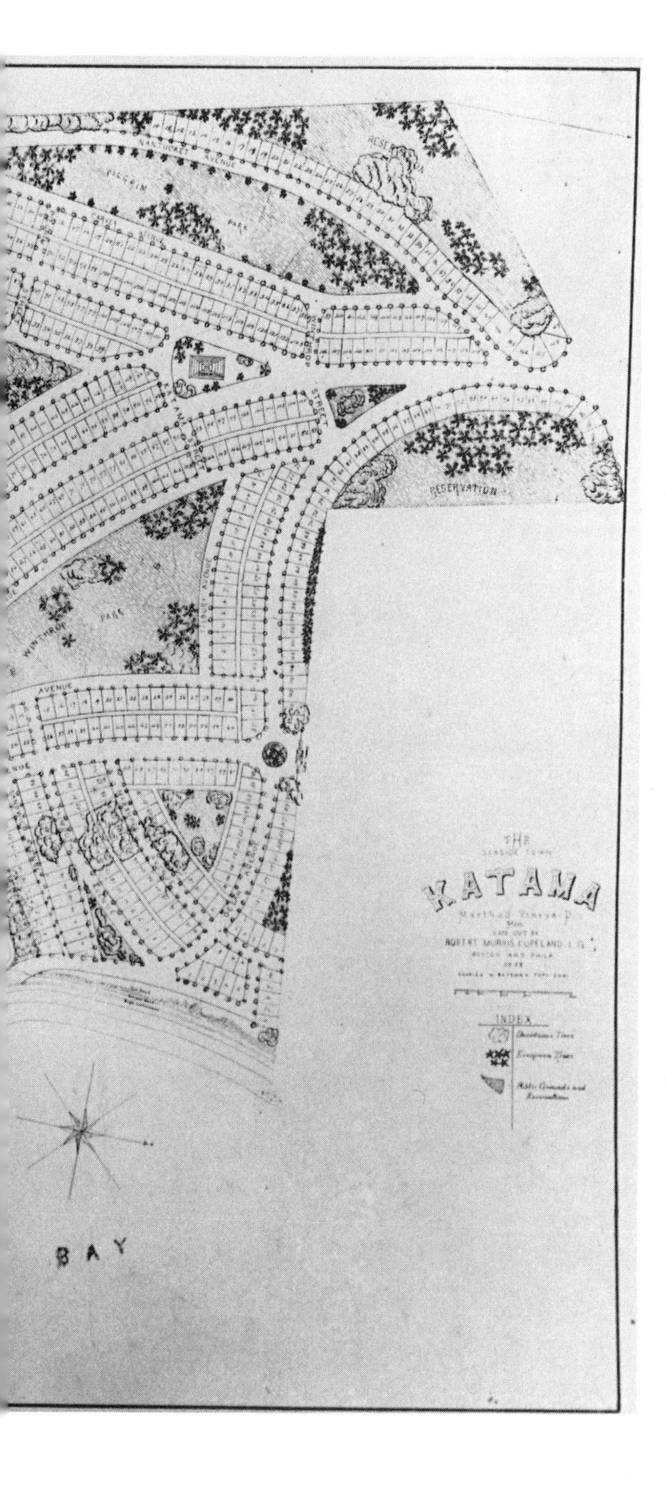
NANTUCKET AVENUE
PARK
STREET
RESERVATION
AVENUE
THE
KATAMA
INDEX
BAY

Round arch cottage in Oak Bluffs. (Dukes County Historical Society/Edith Blake)

included metallic inventions. Other than being the son of a carpenter, we know of no specific architectural preparation. His midlife career as designer was short, including only the work at Oak Bluffs, his own house at Newport (1872), and hotels for two other E. P. Carpenter resorts with Robert Morris Copeland plans—Katama on the Vineyard (1872), and Dering Harbor on Shelter Island, New York (1872). With Dering Harbor came, as well, a Copeland-designed camp meeting for the Shelter Island Grove Association. Both communities were built. A dramatic hotel, the Manhanset House built

(*Opposite*) A. J. Bicknell, *Detail, Cottage and Constructive Architecture,* 1873, a design catalog, picked up S. F. Pratt's designs for Oak Bluffs, plus one poorly drawn campground cottage. (From A. J. Bicknell and William T. Comstock, *Victorian Architecture, Two Pattern Books,* courtesy of American Life Books)

at Dering Harbor in 1872–1873 and destroyed in 1910, completed this direct legacy of Wesleyan Grove and Oak Bluffs,[14] and seems to be the last of Pratt's architectural activities. He became wealthy on a sewing machine patent, and although he was to live on for four more decades in Newport, it was only as a gentleman of leisure and a yachtsman that he was remembered.

Pratt's contribution to Oak Bluffs was that of a dynamic and festive architectural style, one derived directly or indirectly from French sixteenth-century secular buildings as they were then being revived for villas in the Paris suburbs or the Normandy coast. It is also close to the work that Richard Morris Hunt was doing in Newport and to what is known in this country as the Stick Style. It is a hard-edged mode of activated skylines with jerkinhead gables, candlesnuffers, steep-hipped roofs, finials, dormers, and eaves that kick out, bending roof and dormer lines up at their edges in a lilting fashion. As executed by Pratt, the style also has a strong geometric order through centralizing treatment of volumes and a continuous scaling of triangles in elevation, from dormers to windows to gables to the outline of the entire building. It is this volumetric quality coupled with the near-abstraction of the triangles that makes the work so appealing to the modern observer. The flamboyant dimension of Pratt's mode made a visual climax for Oak Bluffs which was more fashionable than the base layer of solemn campground cottages. Aspects of his approach were sufficiently imitated by other builders to give Oak Bluffs a look more reminiscent of Port-au-Prince in Haiti (where French architects were also revamping the continental masonry mode in wood) than of other American resorts.[15]

There may have been 18 cottages designed by Pratt for the campaign of 1870–1872, a number that E. P. Carpenter was reported to have ordered.[16] Only 12 have been located, not all recognizable. Eight of the 12 are of a single type, a nearly cubic mass with a mansard roof, central entrance with second-floor projecting balcony above, and a balcony roof of some kind above that, the whole like a French chateau frontispiece or pavilion. Each had a different form of roof or tower terminating the central entrance pavilion, making the series an exercise in varieties of architectural shapes. Another Pratt cottage type was the Swiss. There were two slightly differing versions of these, both with low wide gables with elaborate fretwork infill. Other Pratt houses have still different forms. One, asymmetric in its massing, with an interesting play of sloping roof planes, was erected for Tillinghast, a Quaker children's preacher from New Bedford. "Round Crown," a name given by a modern admirer, had an octagonal entry unit with octagonal roof, lantern, and cresting. It was built for Hiram Blood,

(*Top*) Mansard-roofed cottage, S. F. Pratt, architect, 1871–1873. (Ellen Weiss)

(*Bottom*) Swiss cottage, S. F. Pratt, architect, 1871–1872. (Dukes County Historical Society/Edith Blake)

Fitchburg, Massachusetts, railway magnate who coined the tag "The Cottage City of America" for the resort.

The most memorable of all Pratt cottages was the one built in 1871–1872 for the governor of Massachusetts, William A. Claflin. The governor was an ardent Methodist who was involved with liberal causes such as abolition and Indian and female enfranchisement. He was the son of Lee Claflin, a famed Methodist success story—penniless orphan as a child and shoe industry millionaire as an adult.[17] The Claflin family arrival at the cottage was an event in the life of the community, and parading the new plank walk past the bath arbor to see the gubernatorial villa overlooking the sea was a season's pleasure. Its powerful forms at tiny scale make it a minor monument of American Victorian architecture. Unlike much design of its period, vertical and horizontal forces are balanced, an anticipation of the Queen Anne and Shingle Style reforms of a decade later. The building is of its era in its sense of hard-edged, internalized tensions, or at least of what reads as tensions to the twentieth-century observer. The easy melodic joys of the popular music of the period suggest that modern interpreters may themselves be too anxious to be able to receive a message of uncomplicated gaiety. The tension, however, to go back to this perception, is between the powerful centripetal force of the cottage's basic geometry with its centralizing density, and the centrifugal projections of strongly modeled, sharply cut, roof and balcony edges. Candlesnuffer, gambrel, turret, dormers, and remaining roof are interlocked in a dense three-dimensional geometric puzzle. From this powerful unit, fine needlelike cresting and finials reach up, like tentative sensors or the cilia of a microscopic organism. Pratt's own home, on Bellevue Avenue in Newport, was a tile-hung version of the Claflin design. Restored and now used as an office, it stands today on its then-very-expensive and still highly visible site opposite the Redwood Library.[18] The Claflin cottage, incorporated into a larger building on the same site, is almost completely lost.

Newport, Pratt's home after 1872, must be seen as the immediate source of Pratt's style, if not France directly. (It is tempting to imagine S. F. Pratt, machinist-inventor, traveling to the Paris Exposition of 1867 and learning this Second Empire mode at first hand.) The mansard-with-frontispiece, the Swiss, and the Claflin houses can all be viewed as witty manipulations of Newport "cottages" of the 1850s and 1860s, those by Richard Morris Hunt, Leopold Eidlitz, and George Champlin Mason, back to true cottage size. Pratt's plastic sensibility probably owes as much to Mason as to the Paris-educated Hunt. Mason's house for Mrs. Loring Andrews, with its push-pull plasticity—Scully's "exacerbated baroque of the picturesque"—is remarkably close to the Claflin design.[19]

(*Top*) H. A. Blood cottage, "Round Crown," S. F. Pratt, architect, 1871–1872. (Vineyard Vignettes)

(*Bottom*) Governor Claflin cottage. S. F. Pratt, architect, 1870–1872. (Dukes County Historical Society/Edith Blake)

It is in the sculptural force, the cross-axial organization, the hexagonal unit with entrance (and dormer above) on the long axis, this contrasted with the gambrel, porch, and secondary entrance on the short axis, and the way the whole pulsating mass is bound together by horizontal bands of sawn boarding that the Newport mansion and Oak Bluffs's cottage show some mutual knowledge. Pratt has simply eliminated the second floor of the larger building while keeping the roof-level complexities, an overblown sculptural top to a one-story structure. The small version becomes a humorous commentary on the larger one, rather than its mere miniaturization, and Pratt has made manifest the popular view of Oak Bluffs as a middle-class Newport.

The two largest elements in the 1870–1872 company building campaign, both designed by Pratt, were the octagonal Union Chapel, which stands today, altered but impressive, and the vast Sea View hotel, which does not. The chapel must be seen as a component of the continuing dialogue between the developers and the camp meeting, a dialogue of both agreement and dissension, with compromise and growth the result of mutual sensitivity and a conscious knitting together of purpose and mood.

No chapel was indicated on the 1866 and 1867 Copeland plans. It first appeared on the February 1870 plan as both an elevation drawing and as a designated place, Chapel Hill, a minor mound behind the apex of Ocean Park, close to the campground. This late offering of an interdenominational meetinghouse could have been a good faith showing of high mindedness or, as one contemporary thought, a way of assuring prospective lot owners access to Sunday services without having to storm the new campground fence and gates.

Union Chapel must also be seen in dialogue with the camp meeting's own new structure. The problem of the dying oaks that sheltered the audience had remained. The old trees were cut down in 1870 and a vast, expensive canvas tent was raised on three poles over the preaching area. The "canvas tabernacle," along with the fence, signaled a deeper commitment of the revival to its site, setting off a flurry of cottage construction within the grove, and further binding the revivalists and the developers to each other. In August 1871, the camp-meeting directors were invited to participate in Union Chapel's dedication. There they heard their own Old Testament imagery, out of camp-meeting justifications of two decades before, as Union Chapel was likened to the "tabernacle of the Israelites, constructed by voluntary offerings of a wandering people as a temporary place of worship." One hopes that the humor of the Exodus metaphor for the summering middle classes was intentional. The richness of ornament of the new building was evidence that the voluntary offerings were heartfelt.

"Those who come here for pleasure or profit have by this building signified their desire that their pursuits shall be sanctified by the Gospel."[20] This peaceful union of the two communities was made visible by other means. Flags of red crosses on white fields (Saint Andrew or Saint George) flew from the highest mast of the canvas tabernacle and from the spire of Union Chapel. Both waved above the trees, signs of Christian unanimity for the passing shipping trade and arriving steamers.

Union Chapel today is an attractive building even with the "colonial" shingles laid over the vertical boarding and white shutters on windows, which have lost the original triangulation of their tops. The triangular massing of the whole structure was made explicit by triangular dormers, ogive windows in the clerestory, and triangular lights over the four corner entrances. These doorways are protected by roofs of two flat gables which intersect to create a form with modernist abstract folded paper qualities. The separate bell tower, with its steep kicked-eaves roof, was built a year after the chapel and is now gone. The complexity of the chapel's planarity,

Union Chapel. S. F. Pratt, architect, 1870. (Dukes County Historical Society/Edith Blake)

its volumetric force, and its crispness suggest something other than French sources, possibly Scandinavian or even Slavic. Pratt's control of form in space does not falter in the interior, where the drama is inverted to space in form. The octagonal, balcony-rimmed, central area, clear up to the base of the spire and delineated by sharply cut planes in easy and apparently weightless assemblage, is impressive by any standards. The interior of Union Chapel stands as an argument that S. F. Pratt, had he continued working, could have been a major American architect.

The Sea View hotel, Oak Bluffs's largest and most visible building was erected on the site of the first company building at the head of the wharf from April to July 1872. In plan, the Sea View was a long rectangle, not as compact or centralizing as the rest of Pratt's work. It was 225 feet long, five stories high on the water side (including the story under the mansard roof), and four stories on the land side. There was a "French roof" on the south

Sea View hotel. S. F. Pratt, architect, 1870–1872. (Dukes County Historical Society/Edith Blake)

end and a "Flemish tower" on the north, the two towers being 85 and 100 feet high respectively. A 26-foot flagstaff topped off the north tower.

In equipment and furnishings the Sea View ranked with the most advanced hotel design of the period, and its luxury and mechanical wonders showed a sharp contrast to the primitive simplicity of the tented camp meeting of a decade before. One could enter the basement of the building directly from the wharf and find a barbershop, billiard room, bakery, icehouse, and space for bowling alleys. Equipment included steam heat, a steam passenger elevator, laundry machines, a water pump, a 2,000-gallon water tank on the roof, speaking tubes, and a Walworth gas machine for lighting the entire building. On the principal floor were reception rooms, private dining rooms, and a 77-foot-long dining salon. The south end of the second story was occupied by a large parlor. The rest of the building held 125 sleeping rooms and private lounges. The hotel was painted in various shades of brown, with the roof finished in "gay colors." One admirer noted that from the sea it looked like an immense private chateau, the seat of a rich man.

The opening festivities for the hotel were a demonstration of the full promotional style of the Oak Bluffs company—500 invited guests including almost everyone of any political profile in Boston and the entire state senate. The inaugural dinner filled the dining salon and the main veranda. The next day the entire party was taken to Edgartown and then to Katama on the south shore to see E. P. Carpenter's newest resort, with plans by Copeland and a combination gateway-and-hotel by Pratt. During its 20-year life, the Sea View housed Alexander Graham Bell, James Gordon Bennett, Oliver Wendell Holmes, President Grant, governors, generals, and titled Europeans. It burned to the ground in September 1892, providentially on a day when Carpenter was visiting, but according to Henry Beetle Hough, a boy about town a decade later, "Its towers linger in all the memories of the generation which knew it."[21]

The collective airy fantasy of Oak Bluffs, with gravity-defying displays of woodworkers' art growing out of the base of odd and solemn campground cottages, was due to still other remarkable buildings not designed by Pratt. There was the Crystal Palace, its name from the almost completely glazed first-story front on which the rest of the building, a Victorian pile of bold bits and patterns, rested. The theme of mass over void, with increasing complexity as the building rose to the roofline, culminated on this building with a steeply pitched, hipped-roof tower set over a porch—that is, with empty veranda space and one prominent post seeming to hold it up. Towers over porch voids were added to many cottages and became an Oak Bluffs motif. The Crystal Palace was built in 1869 in the campground and later

View from boardwalk "over Jordan" to Sea View hotel, campground to right. (Dukes County Historical Society/Edith Blake)

(*Opposite top*) H. C. Clark cottage, the Crystal Palace, 1869, while in campground before the move to Oak Bluffs. (Dukes County Historical Society/Edith Blake)

(*Opposite bottom*) Dr. H. A. Tucker cottage, Hartwell and Swasey, architects, 1872. (Dukes County Historical Society/Edith Blake)

moved to Hartford Park. Its first owner was Henry C. Clark, a Providence coal magnate whose barrage of correspondence to *The Providence Journal* was so opinionated that the newspaper had to devise a new feature, the letters-to-the-editor column, to print it.

Most of the major contributors to the Oak Bluffs skyline faced onto Ocean Park, the better to be seen and to see from. At the apex of the park was the house of patent medicine magnate Dr. Harrison A. Tucker, the resort's unofficial host, whose arrival and departure opened and closed the season. Tucker had offices and homes in Brooklyn, Providence, Boston, and Foxboro, but it was to his home in Oak Bluffs, which he was occupying half of each year at the end of his life, that he came to die.

Dr. Tucker came to Wesleyan Grove soon after the Civil War and was one of the first to buy a lot and build a cottage in Oak Bluffs, choosing a prime site facing the sea. His cottage, the second on the site, was built in 1872 to plans by Hartwell and Swasey of Boston for the princely sum of $11,000.[22] Widely published in the nineteenth century and reasonably well preserved today, it is a rambling building by local standards, with a recessed central block and two forward-projecting wings, all two stories high. The north wing had a third-story "pagoda," a roofed pavilion rising above the trees with walls, a lacy screen of richly ornamental jigsaw trim. The gingerbread on the pagoda and on many porch railings showed animal motifs—griffins, birds, and dogs. The original coloring of the Tucker cottage is not known, but all the board dividers between fields of jigsaw work and board edging of wall panels were painted a darker shade than the rest of the woodwork. Construction was the usual Wesleyan Grove plank frame.

Five years later, in 1877, the Tucker cottage received a round of improvements, carrying it stylistically from a Stick Style to a Queen Anne phase. John S. Hammond of Mattapoisett, Massachusetts, was the architect of the additions. More porches were added and those that were there enlarged. A gable-roofed place-of-appearances was inserted into a second-floor balcony, perhaps in case President Grant returned. Grant had watched fireworks in Ocean Park from that balcony three years earlier. The whole was resurfaced in shaped shingles and clapboards, these covering the tongue-and-groove boarding and dark edging boards which had made the building seem so linear and hard-edged. The remodeled cottage was painted in Queen Anne colors: dark green, Indian red, Quaker drab, and bronze green. Notes on interior colors also survive: terra-cotta, light olive, robin's egg blue, Indian red, salmon, Pompeiian red, and gilt.

The animals in the fretwork of the Tucker cottage occur on a few other buildings in Wesleyan Grove and Oak Bluffs. Two cottages in the campground have them, as does that of Dr. Tucker's neighbor, George Landers,

Dr. H. A. Tucker cottage after 1877 renovations. (Dukes County Historical Society/Edith Blake)

a Hartford manufacturer of furniture, tableware, and metal building parts. Landers's house is in excellent shape today, despite the loss of the unusual octagonal mansard cupola on its tower. The animals on the porches and in the projecting front gable include bird dogs in point, birds with wings outstretched (as if startled into flight), rabbits, and the heath hen, a species that became extinct on Martha's Vineyard in the 1930s. One campground cottage makes the hunting motif explicit: There is a hunter with a gun shown in the vergeboard, the gun being aimed at a bird at the apex of the gable. A fretwork dog similar to that on the Landers cottage is at the Dukes County Historical Society and is attributed to Ezekial Matthews, a builder who came to Oak Bluffs about the time of the 1870s boom. Other sources ascribe animal jigsaw work to the partnership of Samuel U. King and Fred Luce in Eastville, who owned steam-powered lathes for turning posts for the porches that were beginning to be added to the cottages.[23]

Another remarkable edifice stood at the southern end of Ocean Park. This was the Oak Bluffs Club, a cottage-scale social expression of the town's

(*Top*) Landers Cottage, facing onto Ocean Park, 1870s. (Dukes County Historical Society/Edith Blake)

(*Bottom*) Landers Cottage, 1980s. (Edith Blake)

Trim on campground cottage. (Edith Blake)

leadership and, as well, island headquarters of the New York Yacht Club. Likened to the Newport Casino, just as Dr. Tucker was likened to socialite editor James Gordon Bennett, the club combined two existing buildings, elaborated campground cottages arrayed with mansard-roofed towers, peaked dormers, several patterns of shaped shingles, and a Tuscan veranda encircling the whole. Designer of the remodeling, which turned the cottages into an institution, was a summer resident who was the leading architect of Troy, New York, Marcus Fayette Cummings. Exterior colors, as reported in 1888, were dark green roof, Indian red second floor, Quaker green drab first floor, bronze green trimmings, and brown piazza flooring. One parlor was decorated in green, turquoise, cherry maroon, and terra-cotta. A banquet room was in four shades of yellow green. The ladies' parlor was gray and scarlet, and the card room and billiard room pink, Indian red,

Oak Bluffs Club and two other cottages facing onto Ocean Park. Marcus Fayette Cummings, architect of club renovations. (Dukes County Historical Society/Edith Blake)

and scarlet. Unlike Wesleyan Grove in the 1860s, numerous notes about specific hues survive from the 1880s. The Sea View Hotel was repainted at the beginning of the decade from its original varied brown tones to ripe olive with olive green trim and a Venetian red roof. The Joseph Spinney (formerly Isaac Rich) cottage was "fawn" with darker trim striped with plum, and gold leaf on the towers.[24]

The many smaller cottages of Oak Bluffs, which also deserve mention for their originality and verve, are overshadowed by one last large one. Oliver Ames's house was built in 1879 on the bluffs near the Claflin cottage and eclipsed Tucker's in size and splendor. Oliver Ames, son of Oakes Ames, took over the family shovel company after the Credit Mobilier scandal and his father's death, paying all debts, and funding his father's paper legacy of a million to philanthropy. The son was one of Massachusetts's better governors of the 1880s and a cultured and generous man who cherished most of all, in his retirement, the presidency of the Boston Art Club.

With his brother, he was Henry Hobson Richardson's client for the Oakes Ames Memorial Hall in North Easton. The designer of his Vineyard home is not known, but it was splendid with its gingerbread belvedere poised on a tower and its "Moorish" cusped arches edging the verandas. Its present condition warrants restoration. Near the Ames house stood another large one, belonging to A. S. Barnes, founder of the publishing company.

* ** *

THE CREST OF NEW CONSTRUCTION in Oak Bluffs passed by the end of the 1880s, but continuing work on the cottages was as important as any other activity in the bustling life of the town, well past the turn of the century. Each new porch, roof, dormer, shingling, wing, bolstering of foundation, turning of building on its site, or moving it to a new site altogether was avidly reported, together with residents' arrivals and departures or social events. Thus we know that, in addition to the building teams of Wesleyan Grove, other island and New Bedford carpenters were busy at the resort. Some of these were William May, William Chadwick, Lewis and Frank Smith, Alvin Stewart, J. G. Chase, Sam Trippe, William Walker, and Obed Baker. Alden J. Burgess, Ezekiel H. Matthews, Elihu Mosher, and John Alden came from more distant Massachusetts towns. George Rice and the brothers Eli and Barnaby B. Leighton were from Maine. Most became permanent residents, branching out into real estate and other businesses, getting into politics, and finishing their lives on the island. They, partnered with their clients, some thoughtful and visionary developers, and two professional designers, succeeded in creating *de novo* a towered fantasy city by the sea, for everyone's pleasure.

SIX

TWO COMMUNITIES—THE 1870S

READING THE MOOD OR MENTAL TEMPER of Wesleyan Grove in the 1870s, now with its sibling Oak Bluffs, is difficult. Hebron Vincent was no longer logging the annual physical and spiritual accomplishments of his beloved camp meeting, leaving us without his long view and insider's knowledge. Newspaper correspondents' and travelers' pieces got at the externals. Growth, all through the 1870s as in the decade before, was a constant theme. Cottages were built at about the same rate as in the 1860s but, probably because the proportional impact of the new ones on the whole was smaller, journalists no longer attempted an annual new building count. Scattered figures do give some indication. There were 203 cottages and 330 tents in the campground in 1868. This changed to 250 cottages, with only 300 tents, the following year. In 1874 more wood than cloth was reported for the first time—362 cottages and 233 tents. At the end of the decade, a bird's-eye view was taken of what was now known as "Cottage City," which showed 500 cottages and 110 tents in the campground, and 232 cottages in Oak Bluffs. The resort never did outstrip the camp meeting in numbers of cottages, but the houses of Oak Bluffs were larger and many more of them held lodgers. By 1914, Wesleyan Grove was down to 340 cottages, six hotels, and one tent. Cottages could easily be moved to other parts of the Vineyard and to Cape Cod, accounting for some of the loss. Oak Bluffs had 278 cottages and 13 hotels in 1914.[1]

Continued expansion of the campground in the 1870s also meant more commercial facilities. In 1874 there were, within the campground fence, three grocery stores, two meat markets, eight lodging houses, two bakers, two fruit shops, two fancy goods stores, two hardware stores, three photographers, and one each of the following: post office, express baggage office, straw house, association building, and preachers' stand. Most of the commercial structures were grouped into two zones, one at the entrance from the north near the present Wesley House, and one in Montgomery Square, the triangular area onto which Oak Bluffs had opened its arcade. In 1873 the campground put in a horse railway with open cars, the tracks running from the Highland wharf across the pond ("over Jordan") and around the main circle. These were the tracks that would help in the construction of the iron tabernacle at the end of the decade. The shops in the campground may explain the lack of commercial provisions in the Copeland plans, the company assuming that the Arcade and the campground would suffice. Nevertheless, and without change in the deed restrictions, Circuit Avenue quickly acquired stores. It was described in 1876 as having the qualities of a "grand boulevard," and a tourist guide a few years later compared it to the main street of any thriving city with its gas lighting, stores, hotels, and surging crowds moving on and off hotel piazzas and in and out of shell stores.[2] The particularly festive spirit of this street was recorded in photographs well into the twentieth century. Today, with no commercial activity in the campground but also no shopping malls, Circuit Avenue is a thriving business street and a social spine of the town.

Energy for the growth of two communities in the new decade spilled over to fuel a third. The Vineyard Highlands was laid out in 1868 by the Vineyard Grove Company, individuals closely associated with the camp meeting. "Vineyard Grove" was an alternative name for Wesleyan Grove, its postal designation. The Highlands occupied the hills north of the camp meeting and pond and had a plan by a Taunton, Massachusetts, surveyor in a somewhat frozen version of the curvilinear style of its Vineyard predecessors. The core was a circular park, designated for camp meetings if the revival decided to distance itself further from Oak Bluffs. Curving streets surrounded the circle along with a new motif for the island, triangular parks (apex opposite the sea and the shoreline as base) with roads and lots running from apex to base. With the roads and houses fanning out toward the open end at the bluffs, each cottage had an ocean view. Copeland would use this

Circuit Avenue, the "grand boulevard" of Oak Bluffs and commercial center of the two communities. (Dukes County Historical Society/Edith Blake)

device in his 1872 plan for Katama.[3] After 1875, the Highlands circle was regularly leased by Baptists for their annual summer gathering. In 1877 they built an open wooded octagonal tabernacle, now lost. It had a diameter of 135 feet and was 38 feet high at the center with, above that, a 20-foot cupola with a louvred drum and a mansard roof supporting a 93-foot flagstaff. The building seated 2,500 with room for 100 on the platform, and probably inspired the construction of the campground's tabernacle two years later.[4] The architecture enthusiast will enjoy the possibility of the boy

Frank Lloyd Wright wandering about the Highlands. His Baptist preacher father was located in nearby Pawtucket during this meeting's early years.

Development in the Highlands went quickly. By 1870 there was not only the Highland wharf, the camp-meeting's designated landing, but also a plank walk and a hotel. Governor Claflin stayed at the Highlands House before his Oak Bluffs cottage was built. Island carpenters and farmers were among those investing in lots, along with ministers and a famous Methodist millionaire. Isaac Rich was a fatherless child who sold fish from a barrow on the Boston wharves to support his ailing mother. He made his fortune in the fish trade and became a principal founder, with Lee Claflin, of Methodist Boston University. Grieved by the death of all of his children, he turned his fortune to philanthropy, giving a million and a half dollars to the school, which he refused to have named for himself.[5] Isaac Rich's towered villa in the Highlands, next to more modest efforts owned by humbler coreligionists, suggests that the idea of being middle class had more to do with social, religious, aesthetic, and recreational values than with actual economic achievement.

The growth of Wesleyan Grove, Oak Bluffs, and the Highlands during the 1870s was climaxed at the end of the decade by the creation of a new incorporated Massachusetts town, Cottage City. Edgartown did not want to lose the booming resort's taxes but was not providing fair services, according to the secessionists, and the battle for independence was bitter. Among the published arguments for division was one describing the three areas in February 1879. There were 767 cottages, 1,058 taxable buildings, two steam mills, two schools, 15 hotels, two markets, two blacksmiths, seven grocery stores, and three churches.[6] The schools make explicit a new dimension for the revival-resort, a winter population. That month 150 families were in residence, with 132 voters. Many of these year-rounders were carpenters, others newly arrived farmers and fishermen from the Azores. There were some winter residents at least eight years earlier, for in December 1871, in the campground, Wesley Grove Vincent was born.

Even in the 1870s, as Wesleyan Grove and Oak Bluffs pushed to the point where they could be a town, the place continued to amaze its visitors. Themes of astonishment, of having been transported from the mundane world into a fairyland, the shock of a miniature city dedicated to joy, pervasive religious feeling, nature, and social density permeated the new decade, even with the resort's blatantly commercial edge. Journalists were now writing from the press room in a tower of the Oak Bluffs wharf gatehouse, commanding a view of the crowds gathered to watch steamboats jostling for a place at the pier. The press now repeated fewer complaints about religion being subservient to fashion and the "pic-nic spirit," presumably because those of more ascetic temperament had given up. The general gaiety made simultaneous religious and financial profiteering acceptable. The observation that, because the campground gates were now being left open all night, a visitor could go directly from evening religious services to the sale of lots, was recorded with warmth and humor. In early August 1871 the scene was "swelling in numbers and excitement," with "happiness more and more contagious and business thriving."[7] At noon, the accustomed time for bathing, 200 swimmers were seen in the water, men and women together, moving to the rhythms of the Foxboro brass band playing from the bath arbor on the bluffs. The band worked almost continually—for arriving

Wharf gateway, with part of Sea View hotel. S. F. Pratt, architect, 1868–1872. (Dukes County Historical Society/Edith Blake)

boats from the upper balcony of the gatehouse, later from a special pavilion on the wharf, and from the Sea View hotel piazzas. Oak Bluffs seems, in part, matched by the lightness and joy of the Impressionists' views of fashionable watering places on the Normandy coast. Monet's 1870 paintings of the Hôtel des Roches Noires on the beach at Trouville show how the broken skyline, so close to Pratt's Sea View, worked with the scudding white clouds, choppy white-edged sea, bright sun, rippling flags, and colorful dresses and parasols to create a visual dream which would blot out in freshness and beauty any shoddiness in economic underpinnings.

The most vivid descriptive prose was now coming from guidebooks and magazines, the newspaper correspondents presumably having worn out their amazement. The new travelers repeated or reinvented the imagery of the 1860s, showing the resort's success in continuing the sense of the campground. "There is an indescribable charm, an undefined sense of strangeness apparent at every turn." New arrivals were greeted by flying bunting in every direction, the welcoming strains of music from the hotel, bands wafting across the water—all combining to impress the stranger with the idea of a "fairyland." A guidebook tells of the "city of quaint cottages . . . old and yet not, grave, though gay, solemn, yet hilarious." It is where those who fled one city were content to inhabit another. Campground cottage doors were still wide open and families still lived on the piazzas, showing an "airy" existence not to be seen elsewhere. "In the semicircle, hundreds of cottages, minarets, pagodas, an enormous gospel-tent, hotels with bunting floating gayly everywhere, and everything basking in the dazzling sheen of the brightest and whitest of sunlight. Was it a toy? Was it a picture painted in those vivid colors with which the younger school of French art loves?" For this well-traveled voyager, Oak Bluffs outstripped all the strange places of Europe and North Africa and was still unlike anything he had ever known. "We gazed from the deck of the slowing steamer, amused, fascinated, incredulous. Everything looked so small, everything looked so strange, everything looked so bright. Where did the inhabitants reside: Not in those sentry boxes, those card-houses, those Lilliputian edifices."[8]

Many were caught by the peculiar, the odd, the "foreign and bizarre appearance." "It is sort of a Mayfair of pleasure, a city of the night, which is unreal and insubstantial in its beauty and apparently as likely to pass away from sight at any moment as the ships on the water horizon." This is Charles Dudley Warner, popular essayist and Mark Twain's coauthor for *The Gilded Age*. As late as 1886, Warner wrote of the "fantastic tiny cottages among the scrub oaks." Family life was on display in the

Unidentified cottage, Oak Bluffs. (Dukes County Historical Society/Edith Blake)

. . . painted little wooden boxes, whose wide open front doors gave to view the whole domestic economy, including the bed, center table, and melodeon. . . . In the morning this fairy-like settlement, with its flimsy and eccentric architecture took on more the appearance of reality. . . . The foreigner has a considerable opportunity of studying family life, whether he lounges through the narrow, sometimes circular streets by night, when it appears like a fairy encampment, or by daylight, when there is no illusion. It seems to be a point of etiquette to show as much of the interiors as possible, and one can learn something of cooking and bed-making and mending and the art of doing up the back hair.

But Warner the sophisticate is separate from these people. He implies that there was a mechanical tone to the hilariousness following the band from wharf to hotel piazza to bathing pavilion and back. The core quality of the place, he suggests, is "peculiarly Yankee—the staid dissipation of a serious-minded people." "Most of the faces are of a grave, severe type, plain and good, of the sort of people ready to die for a notion."[9]

Themes specific to Wesleyan Grove in the 1850s and 1860s survived all of the resort panoply now encapsulating the revival. One could still read how individual families melded into community, and how the accompanying feelings were an antidote to the selfishness of the age. ACG, the newspaper correspondent who soon would write about Llewellyn Park as a monument of American economic democracy, was struck by the communal temper of Wesleyan Grove. "Everybody lets some of the 'good time' out the wide front door for their neighbor to note and the stranger that is within the gate to look at. Everybody rejoices in what others have, for is it not inevitably in part their own? In this simple out-of-doors life, one could not be selfish and build high fences." Another correspondent, noting the loving greetings of summer neighbors who had been separated all winter, said that this was "good to see in these days of selfish isolation." Even James Jackson Jarves's full Utopian interpretation was still viable. As late as 1880, with resort life thriving, the Reverend Frederic Denison, a Rhode Island minister and accomplished local historian, thought that the camp meeting community represented "a new phase in the world's social history." As for Oak Bluffs, "The wilderness was made glad and the desert blossomed as the rose. . . . Order, prosperity, quiet, fraternity, cheerfulness, joy, and profit. If the secular has been welded to the religious, has the union been a violation of the letter or spirit of Christian law and liberty?" One would like to have the views of the British Christian socialist Thomas Hughes, who, Denison informs us, liked it when he saw it in 1875. Hughes was in New England seeking support for his own effort in community creation, his "New Jerusalem," Rugby, Tennessee. Because he also wrote travel sketches for the British press, his experiences may yet come to light.[10]

With the 1870s and the revival-resort's wider visibility, unfavorable criticism of its visual qualities and the values represented also began to be heard. Geologist Nathaniel Southgate Shaler, on the island in 1872, disliked it thoroughly.

> Oak Bluffs is a mushroom town without any oaks, except some scrubs, and little in the way of bluffs, except what one gets from the superchristianized people. White pine in the shape of gothic shanties is the only forest growth I have yet found. . . . There are hundreds of box-like houses of a queer and profane architecture occupied by people of the middle classes or

> waiting for some one of that class to buy them. These little dabs of dwellings [are] about as big as boardinghouse slices of minced pie. . . . There is no visible kitchen to them, nor any outward means of existence, unless they live on acorns or are fed by the woodchucks or the emaciated crows.

Two years later Shaler placed a piece in the *Atlantic Monthly,* which characterized the place as a "pasteboard summer town capable of giving bad food and an uneasy rest to twenty thousand people." Its inhabitants showed the admirable lack of discrimination characteristic of those who haunt the shore in summer, their ambitious houses overshadowing and blighting the trim little boxes of the sea-faring class.[11] Shaler's tastes are those of an aesthetic conservative and his actions consistent. He purchased an old "sea-faring class" Greek Revival farmhouse in the remote hilly reaches of the Vineyard as a retreat from the rigors of Cambridge academic life. His Seven Gates Farm was later subdivided so that no house could be seen from any other.

Academics, as a group, probably felt the most active distaste for the style of the evangelical mainstream of American Protestantism. Andrew Dickson White, historian, diplomat, and president of Cornell University, saw Oak Bluffs in 1886 as "a sort of saints' rest where, during the summer, a certain class of pious New Englanders of the less intellectual type crowd themselves into little cottages and enjoy a permanent camp meeting." He found the religious content repulsive, reminding him of the whirling dervishes of Cairo. The sight of young people roller skating with arms about each others' waists to a waltz version of "Nearer, My God, to Thee" so offended him that he was moved to charge all camp meetings with blasphemy and eroticism.[12] White's reaction is reminiscent of Frances Trollope's 50 years before and suggests that for all of the institutionalization, architecture, and middle-class triumph, the core spirit of revivalism might be consistent. Those who would have found it repellent early would find it so late.

With the 1870s, the growing revival-resort acquired a new celebration on which journalists could focus. The Illumination was an evening of oriental lanterns, fireworks, music, and parades. Although such festivals were hardly invented at Oak Bluffs, their island phase began in 1869 with presentations by the developers in Ocean Park, and independently or not, two residents of Clinton Avenue in the campground. The success of the Illumination, like that of the cottages, can be measured by numbers. In 1870, 100 lanterns brightened Clinton Avenue. In 1872, 750 lanterns lighted Ocean Park and parts of the campground. In 1873 there were 1,500 lanterns in Clinton and the following year 3,200 lanterns plus two locomotive headlights. Illuminations were held during the week preceding the actual camp meeting, the largest, in Ocean Park, the Saturday night before the

Cottage bedecked for Illumination, Oak Bluffs. (Dukes County Historical Society/Edith Blake)

revival began, furnishing a climax to the frivolous part of the summer. Illuminations evoked wonder for the magic they created and admiration for the strict decorum and quietness of the crowds. When the red, blue, and green lights came on in the campground's Cottage Park, the silence was so complete that the voice of a small child lisping "Three cheers for Mr. Light" was heard all over. Anywhere else, it was noted, there would have been loud huzzas and cheering. Methodist temperance was credited for the calm.[13] Illuminations became the way for the decade to create anew the dazed silence, the sense of utter amazement, the unreality of this exotic place. It was an upgraded and opulent version of the "phantasmagoria of light and beauty" of the lighted tents of the 1850s and of the nighttime theater of the first camp meetings.

Henry Beetle Hough has recreated the spirit of Illuminations as well as anyone will ever be able to do, but one can also look at primary sources. The festival of 1877 was especially admired by the *Providence Daily Journal.* It was a scene of carnival brilliance as never witnessed in the United States, comparable only to Christmas at Rome. Special boats arrived throughout day from all over southern New England, bringing an estimated 30,000 people. The celebration was underway by eight, with cottages decked with lanterns and flags on porch roofs, eaves, pinnacles, and towers. More lanterns were strung across avenues. Hotels were ablaze with lights, and yachts in the harbor had multicolored lanterns hanging from their rigging. Red and blue rockets and fireworks of every description were set off, and it all looked "more like a glimpse of fairyland than of reality." A procession led by the Mansfield band started at the Sea View and marched up and down the avenues of Oak Bluffs, its progress marked from the distance by a continual blaze of light from fireworks and rockets. The American Band of Providence played in front of the campground's Central House and other bands worked in other locales. The spectacle began to wane by ten and was finished by midnight. Many who had come that day walked about all night for lack of rooms, whereas others bedded down on steamer decks, piazzas, the seats of the tabernacle, and billiard tables.[14]

One aspect of Illumination decor is particularly revealing of the mental temper of the participants. Cottages on Clinton Avenue were draped with mottoes, with more messages on banners spanning the avenue. In 1872 there were pro-Grant political slogans and a Horace Greeley caricature, the latter an oak knot with wool "hair" and spectacles, an item that reappeared in successive Cottage Park and Ocean Park Illuminations. Messages were a mix of religious sentiment, summer celebration, hometown boosterism, and punster fun. "With light and song we greet you." "The Vineyard is our resting place, Heaven is our home." "Walk while ye have the light lest darkness come upon you." Individual cottages sported personal notes, often executed in ferns or evergreens: "We trust in Providence, Rhode Island." "We'll camp awhile in the wilderness and then we are going home." "Our home is by the sea, Heaven is beyond." "Sing songs and be glad as you go." "Hurrah for the Nutmeg State" (Connecticut). "Adam's Eden" was echoed by "Eve's Also" on the other side of one door. "The Fairies' Abode" was placed on a cottage occupied by some small ladies. "Ohio greets the Cottage City," "One Year Nearer Home," "New Yorker's Retreat," "A *light* repast we offer you," and "Here the weary ones rest," were also noted. A display in front of Bishop Gilbert Haven's cottage showed an open Bible with a hand pointing to the words "Follow me," this juxtaposed with a box of mineral specimens to show a harmony between geology and Genesis. A neighbor displayed a transparency of a halo of light around a rock with a cross on it.[15]

Many Illuminations had an odd dimension debated at the time for appropriateness but reappearing nevertheless, the Antiques and Horribles Parade. This custom, alive today in rural New England, was a nineteenth-century Fourth of July tradition in Pawtucket but also had to be explained for his readers by a Boston journalist as being like a Mardi Gras parade. That of 1874 was led by a tin-pan drum corps followed by ghosts wielding croquet mallets, Chinese and Japanese figures, Siamese twins, "indescribable things and a Satanic majesty," a comet, a Brigham Young, and a walking caricature of poor Nancy Luce.[16] Miss Luce was a partially deranged island woman who sold to tourists copies of love poems made for her pet chickens. She was visited by Oak Bluffs denizens as an object of mirth or pity, depending on mood. The eerie dimension of the Antiques and Horribles and the religious resort's preoccupation with Nancy Luce suggests that the theater of summer release must include the exercise of a wide range of fundamental human emotions.

For all of the hoopla, bustle, beauty, and strangeness of the Illuminations, it is also apparent that the psychic center of the summer in Wesleyan Grove and Oak Bluffs was its calm, its silence. The religious colony had survived not only the surrounding festivity and financial speculations but had kept itself as a quiet core to the development, working in concert with the empty center of Oak Bluffs, the open vista to the sea. Religious work in Wesleyan Grove now began in early July, with morning and evening prayer meetings and several Sunday services. Life was simple and domestic. Fathers could be with their children all day, a novelty for the businessman. Tired mothers would be refreshed as well, even with husbands gone during the week, for it was safe to let children roam. Fathers could leave wives and daughters without worry about offenses to their sensibilities or persons. Life was a "green and refreshing oasis in the desert of business and conventional life," free from the "low and vulgar influences universal in our land." Ministers commuted as well as businessmen, spending the week on the island and returning home to fill pulpits. The community was said to be so peaceful that it was hard to tell which day was Sunday, and this "Sabbath calm" in turn was called forth to explain why summer residents came from as far as Baltimore and St. Louis. "Life seems one gala day and pleasure, innocent and healthful, seems to be the object of all."[17] There was a softness, a languor, and an ease. Only the clicking of croquet mallets or the

Clinton Avenue, with cottage of Bishop Gilbert Haven, where President Grant slept. (Dukes County Historical Society/Edith Blake)

lapping of water against the shore interrupted thought. It was a place to do nothing, but this "nothing" included the necessary human task of watching others and meditating on the course of life.

ACG (the *Journal* correspondent) brings us in close to the religious center. He began his letter to the city, as so many did, with a description. The island looked as through God's finger touched but did not press. From the arriving boat, the land showed gentle curves with a background of dark trees to the "bright relief" of the summer homes in their infinite variety. The avenues of the town followed nature, without impertinent straight lines. It was in this nest of soft beauty that ACG was spending the morning, motionless on an Ocean Park piazza, watching the sea. (Of the afternoon activities yet to come: "Walking is no weariness and the world of Martha's Vineyard marches in twos and threes to the landing to see the boats come in and hear the band play, after which we take a stroll through the sacred enclosure and are never tired of the glimpses we get of a sweet and social home life.") ACG admired the women of Oak Bluffs, whom he found sensibly dressed with intelligent faces "not marred by small purpose in life." But soon our piazza-based observer turns his attention back to its magnet—the sea—and gives it a meditation: The attraction of the sea lies in its fascinating likeness to our own humanity, for it is sometimes smooth and sometimes stormy. It is sometimes a thing of commerce, "and life for us is one vast bargain and sale," and it is sometimes a thing of pleasure as "white-winged hopes sail gayly out to the horizon of our lives—that meeting place of the future—and clearly divined for a moment only, return to us to sail to

Band pavilion overlooking Nantucket Sound. S. F. Pratt, architect, 1871. (Dukes County Historical Society/Edith Blake)

an unknown shore." But the sea also has rocks on which strong ships are dashed to pieces, deceitful calm, treacherous brightness.

> The passion and the power, do we not know it all? But we climb from lower truth to higher, and remember that He who "holds the waves in the hollow of his hand" has a measure also for the tears of His beloved, and when all of storm and sorrow, all of warmth and beauty has been experienced, and we have struggled as all souls must till well nigh faint and lost, He can speak to sea and soul the everlasting "Peace."[18]

Religious content in an ocean contemplation was probably a convention of the period. More than a decade earlier *Harper*'s "Porte Crayon," David Hunter Strother, wandered the other end of Martha's Vineyard on a fishing holiday. His 1860 essay on the adventure was lighthearted in tone, focusing on nature's charms, the picturesque characters he met, and the quaint ways of rural island life. Sandwiched between amusing encounters with deliberately provocative island folk is a quick trip to the beach at Gay Head and an abrupt shift in mood: "But to steep the soul in awe, to awaken in the mind the highest sense of sublimity, what is there like the oneness of the ocean." The simple horizontal blue, the measured surge of watery power, the solemn anthem of "eternity's unceasing roll" are the imprint of the spirit of God from when he first touched the waters. And this is fixed, immutable, continuing to the end, "when the heavings of this restless bosom shall be stilled, this voice of thundering hushed—on that day when the great closing anthem shall be sung: 'Dies irae, dies illa, saeclem solvet in favilla'." (Strother then examines the beach to see what has been washed up.) The meditation was illustrated by a drawing of a standing figure looking out to sea, a drawing that became the basis of a well-known anonymous primitive painting in the Karolik Collection at the Boston Museum of Fine Arts, suggesting an interpretation for New England seascapes such as those by Kensett, Heade, Ryder, and Homer, as well as a wider context for our understanding of Ocean Park and its vista.[19]

D. H. Strother's and ACG's demonstrations of the religious content of an ocean view go far in explaining how the revivalism of Wesleyan Grove survived Oak Bluffs. For even at the latter, a meditative quiet was the center of all the pageantry and material display, and it was this which anchored the architecture and play. This underlying placidity of life could "joyfully" be put aside for the earnest work of camp meeting. Evangelical religion was still the core of cottage life and the camp meeting the nucleus of the great summer resort, the "modern Bethesda" that Hebron Vincent had foreseen.

Camp-meeting intensity did fade eventually, at a rate that has not been

D. H. Strother ("Porte Crayon"). The beach at Gay Head. (*Harper's New Monthly Magazine,* August 1860)

measured, but accounts of the 1870s show a continuing strength for the form. Those who had been there for 30 years found the meeting of 1869 the most memorable of all with "some of the grandest thoughts ever attained," freedom from criticism, and a spirit of religious catholicity; 10,000 or 12,000 people thronged the grounds several days before the revival, and yet a "Sabbath stillness" prevailed. Governor Claflin addressed a children's meeting (watched by 6,000 to 10,000 adults) to say that Sunday schools, as inculcators of Christian values, were the hope of a nation beset with immigrants. The Hutchinson Family Singers, frequent visitors over the years, collected funds for the Vineyard Church in Hutchinson, Minnesota. And Hebron Vincent, closing his history, wrote about the human need for Methodist teachings. "The mass of the people cannot contend successfully with the cavils of infidels like Hume and Voltaire and others, but they can tell what they feel and know, which is a stronger argument than any mere deduction of the intellect. . . . The great object here is to know Jesus. And the great inquiry of the saved in heaven will be to see Jesus." When the meeting closed, all were overwhelmed by a powerful sense of the presence

Revival meeting under canvas tabernacle, 1870s. (Dukes County Historical Society/Edith Blake)

of the Master, and "the great sense of Union in the vast prairies of Christian union."[20]

The spiritual strength of 1869 suggested to some a revival of the doctrine of entire sanctification, or Holiness, the Wesleyan perfectionist goal which had fallen into disfavor. Fervor was high and a proposal was put forth to hold a National Camp Meeting Association Holiness meeting the following year. This group had been organized two years before in Vineland, New Jersey, to purify and revitalize revivals to their pre-Civil War, pre-resort form and to restore Wesleyan perfectionism. Nothing came of the idea for the Vineyard, although Holiness camp meetings were held at other sites for several decades.[21] There was a second Holiness flurry at Wesleyan Grove after the meeting of 1871, a minister who was a member of the National Camp Meeting Association holding supplementary sessions. But the resolution to stay and finish the work failed as the remaining worshipers' spirits were felled by the loneliness and the bare tent frames of the abandoned community. Many times enthusiasts vowed to stay for a September season but left early, overwhelmed by the barrenness of the grove without its crowds and by the sense that life had moved on elsewhere.

It is hard to gauge the religious achievements of the meetings of the 1870s without Hebron Vincent's summaries. Newspaper accounts are lengthy but unsubstantial, listing speakers, texts, sermon titles, and the occasional sermon summary. There are some descriptions in dispersed contexts by outsiders, not all of them sympathetic. Teenaged Henrietta Hawes wrote in her diary in 1872 about a trip to the Vineyard and her mixed feelings about the revival. She had come with her family and some girl friends by train from central Massachusetts and was staying at the Sea View. "I liked the prayers, they seemed so earnest and the people begged so hard for what they wanted, but it would have been as well if two or three had not tried to pray at the same time. . . . There was something very impressing and touching about it all, yet I could not help feeling that in some cases it might only prove mere excitement and that perhaps the religion would be all gone to-morrow."[22] The girls retreated earlier than intended to their hotel rooms so that they could laugh without offending anyone.

The visit of President Grant in 1874 provided another glimpse of continuing revivalism. Bishop Gilbert Haven preached his famous sermon, "Multitudes, Multitudes," and afterward the Reverend L. D. Bates conducted an old-fashioned exhortation for sinners to come to the altar. Grant was seated on the stand.

> The truth was hot, pungent, convincing, and convicting. At the close of the sermon [Rev. Bates] invited sinners to the altar, and amid weeping they came from many parts of that great congregation of at least ten thousand people. Men of less faith were anxious about the General, how he would take it. Mrs. Grant was in tears. The Bishop went down into the straw to point sinners to Christ. The President's wife was on her knees, her husband sat with folded arms looking down upon the scene. One of the polite managers asked him if he did not wish to retire from the stand. Said Grant, "No I propose to stay and see this thing out." And he did. Victory turned on Zion's side and amid the shouts of many redeemed souls, the meeting closed.[23]

Harriet Beecher Stowe was present in 1875 and praised the religious tone. In the spirit of B. W. Gorham 30 years before, she argued that, as one learns from the ancient Israelites, religious feelings can be helped by variety and pleasant rural surroundings. "It may be a sign of better days that cars and boats run not to places of dissipation but of worship."[24] The meeting of 1876 was a "largest ever" situation. More were attending the first session on Sunday than any of the earlier festivities, even the Illumination. Camp meeting was called the soul of cottage life and the multitudes of people roaming the area were deeply imbued with religious feeling. Still, something was being lost. Despite the continuing religiosity, newspapers no longer recorded the amazements of ordinary people. The lively debates of the cottage-building era of the sixties over the interlocking issues of growth, materialism, and aesthetic value disappeared. It had become an old story. Thus the two communities' two largest structures, both dedicated in July 1879, came into being with almost no attention to architecture as a public issue. The short-lived roller-skating rink and the permanent iron tabernacle, today Wesleyan Grove's grandest building, received such bare notice that it is hard to piece together even an outline of their history.

Oak Bluff's roller-skating rink, built by skate magnate Samuel Winslow of Worcester, Massachusetts, was an early foray into a specialized building type. Placed just north of the Sea View hotel, it was damaged in the fire of 1892 and torn down. The rink was a huge building, 184 by 87 feet in plan with walls 37 feet high, and was modeled, apparently, on city armories. It was spanned by 70 structural arches made of spruce planks bolted together, these supporting an iron roof. On the ocean side sections of wall could be rolled aside so that skaters could seem to move with the sea. The exterior was decorated in false Stick Style half-timbering, and it had a bevy of turrets and cupolas, all topped by flags, at least in the promotional illustrations. It was the scene of many gay festivities illuminated by oriental lanterns strung through the cavernous space. Here young couples skated to a waltz version of "Nearer, My God, to Thee," to the disgust of Andrew D. White.[25]

The iron tabernacle, which was also slipped into the complex fabric of the two communities without sufficient attention, survives today, with only minor changes, as a remarkable American building. The old canvas tabernacle, erected in 1870 over the traditional preaching area, never worked well. It always had ventilation problems, despite circular gaps in the fabric around the three masts, and in storms it frequently collapsed. In 1874 a proposal was made to build a structure which would combine the need for a chapel for the growing winter population and tabernacle for the summer crowds. By August 1877, this single program had been split into two, and the following spring a substantial chapel seating 250 was erected.[26] The chapel was placed where the County Street, Fourth Street, and Chilmark society

The Iron Tabernacle. John W. Hoyt, engineer, 1879. (Edith Blake)

tents had stood, with its entrance on the inner side of the circle, near the preaching area. Designs were by Edward L. Hyde, a Methodist minister then assigned to Middletown, Rhode Island, who had been trained in art and architecture.[27] The church, an attractive, though not unusual building, is a combination of Stick Style and Queen Anne modes.

By the end of 1878 plans were underway to complete the 1874 building program with a permanent tabernacle which would seat 3,000 to 4,000. Competing designs for wooden structures came from three architects: S. S. Woodcock and E. N. Boyden, both of Boston, and Caleb Hammond and Son of New Bedford. The Woodcock design won, but the 17 construction bids opened in March 1879 were disappointing, ranging from $10,000 to $15,000 for a building budgeted at $7,200. In the middle of April a contract was made with Dwight and Hoyt of Springfield Massachusetts, for an iron tabernacle to be completed by July 1 at a cost of $6,200. The final cost was $7,147.84.[28]

The iron building that the camp meeting association got for half the price of a wooden one is a remarkable structure. Three tiers of roof seem to hover above a nearly circular space about 130 feet in diameter on a north-south axis (taken from the outer supports but excluding the roof overhang.)

This area covers about a quarter more ground than the original preaching area. The stage of the new building was located on the site of the old preaching stand, but the building was "swung" about 40 feet to the southwest, with the stage as pivot, so that the audience area was shifted somewhat in location. This is the only change in the primitive, consecrated meeting ground since 1835.

The roofs, looked at in elevation, show angles that shift slightly, the lowest one being the steepest, suggesting to modern eyes something like a stop motion photograph of the beating wings of a bird or spacecraft with with thin tentacles dropped to the ground. Visible support for the great roofs is a web of fine wrought-iron arches above the thinnest of point supports, except for the area where the wooden stage wall appears to provide more substantial underpinnings. A nineteenth-century observer, criticizing a drawing of the new building made for the association's letterhead for failing to show the stage wall, said that the building without the wooden wall looked as if it had no support, and would fall on its inhabitants. The roofs, originally corrugated iron and now a corrugated asbestos material which retains the original texture, are separated from each other by bands of clerestory windows. The lower clerestory has colored glass windows about 2 feet high, whereas the upper windows are about 8 feet high, suggesting increasing verticality for the higher portions of the form. The top roof is low-hipped, square in plan, and has a wooden cupola that tops out at 100 feet above grade.[29] The lower roofs have progressively rounded corners, so that the building appears nearly circular at base. There is a marked entrance into the open building, an arch cut into a chunky mass like the base of a tower, on the east side opposite the stage.

There are also entrances on the north and south sides, arched openings and gable roofs, but no "tower base." The three entrances were described in the nineteenth century as giving aesthetic relief from the expanse of roof, protecting worshipers from rain dripping from the eaves, and adding "symmetry." The entrances and stage wall fix the building, making front and sides distinct from each other—thus, presumably, the "symmetry."

The structural core of the tabernacle consists of four major iron supports placed about 40 feet from each other, forming a square in plan at the base of the square top roof. These vertical supports are actually trusses, the upper ends arching in toward each other and meeting at the center, about 75 feet above the floor. As the best contemporary description put it, the inner chord of the truss is an arch, whereas the outer chord is a vertical and then an inclined brace. These vertical trusses are fixed rigidly at the level of the upper clerestory by four horizontal trusses, these lying on edge, parallel with and just behind the windows, almost invisible against the light. The horizontal trusses make a square in plan and are the only circumferential metal members attached to each other, metal to metal, in the building. Beyond this square, all metal-to-metal members are radiating lines, the circumferential connections being in wood. This makes the building essentially wood-jointed, allowing for ease in construction on the uneven, sloping ground by eliminating the need for retooling of metal parts on site. The alighting bird or hovering spacecraft is a flexible building machine.

The radiating metal-to-metal structure extending under the two lower roofs consists of delicate arches of 2-1/2 by 2-1/2 inch T-sections which hold radiating or longitudinal iron rafters which in turn hold circumferential wooden purlins. Five arches fan out from each of the four vertical trusses, coming to the ground in a total of 20 supports. From these 20 posts, the arching motion starts again, this time with 2 by 2 inch T-section arches ending at 20 more supports at the perimeter. (There are also 12 extra iron supports grouped about the three entrances, their iron connectors going directly to the horizontal trusses, rather than to the four vertical supports. The connectors are not arches but, rather, iron beams and tension rods.)

The secondary and tertiary vertical supports consist of bunches of standard pieces. The secondary supports are made of four T-sections in-

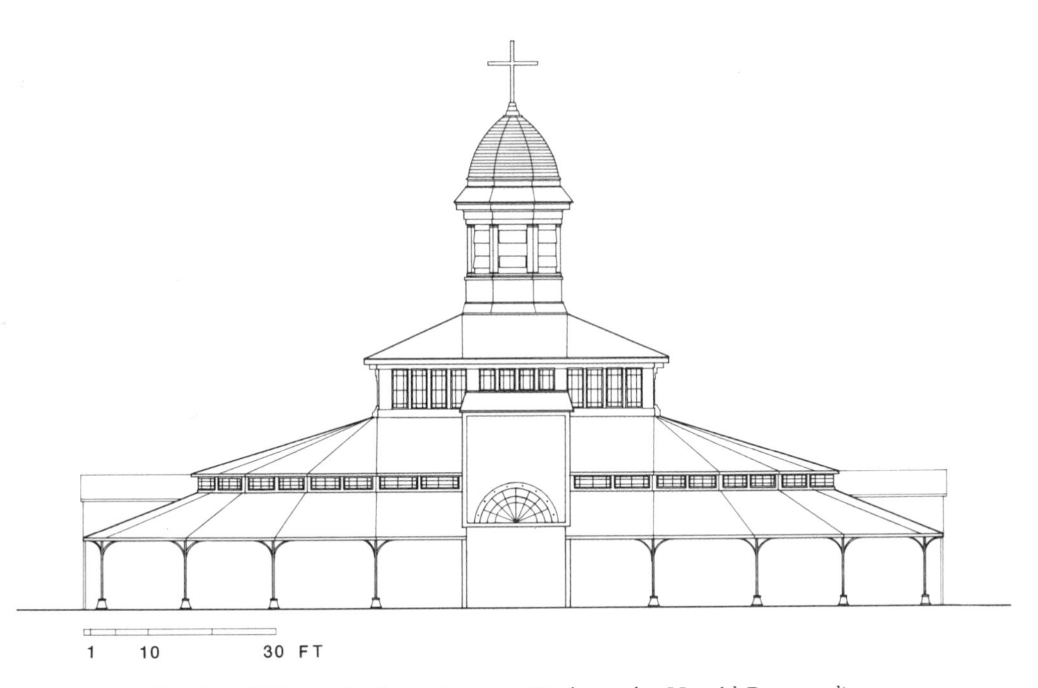

The Iron Tabernacle, from the east. (Redrawn by Harold Raymond)

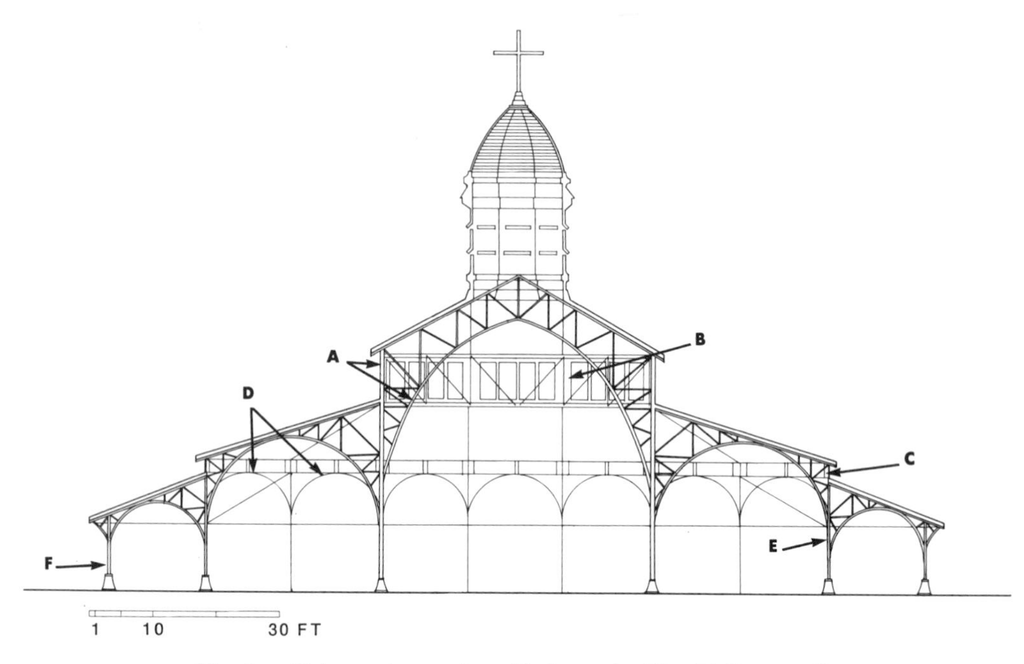

The Iron Tabernacle, section. (Redrawn by Harold Raymond)
A. Verticle truss with arched chord; B. Horizontal truss in front of upper clerestory; C. Lower clerestory; D. Arched brackets; E. Secondary supports; F. Tertiary supports

terlocked, the tertiary supports, of two T-sections and two 2-inch pipes. All posts vary in height, accommodating the uneven slope of the preaching area. This makes the roofline, as seen from outside, level, rather than parallel to the ground, and reinforces the illusion that the roofs are like an animate being, mobile, independent of the site.

The thoughtfulness with which the visual quality of the tabernacle is integrated with its structural and constructional determinants can be illustrated with an interior detail. The lowest roof is encircled at outer and inner edges by thin iron arches spanning the bays between the vertical supports. These arches are actually brackets, overlapping at the seeming crown in some places or not meeting at all, depending on whether they are crossing a wide or narrow bay. They attach to wooden purlins. Thus a place where the irregularity of the building might be most apparent, near eye level, is masked by a device of decorative as well as structural use, giving a lifting effect of lightness and grace, much like the nonstructural arch at the base of the Eiffel Tower in Paris, built 10 years later.

Construction of the tabernacle went quickly and almost without incident. A strike at the rolling mills meant a delay. An accident that sent an iron beam and three men to the ground frightened onlookers but caused no serious injuries. One vertical truss is stamped "Phoenix," indicating the Phoenix Iron Co. in Phoenixville, Pennsylvania. The first iron parts, which had been preassembled in Dwight and Hoyt's Springfield shop, arrived on the island June 9, 1879, less than two months after the contract was signed. The vertical trusses came from Springfield by boat, down the Connecticut River and up Long Island Sound. The horizontal trusses and smaller pieces went by rail from Springfield to Woods Hole, on Cape Cod, thence by steamer to the Highland wharf, and by flatcar on the camp-meeting's new horse railway from the wharf "over Jordan" into the grounds. By early July the frame was finished and the roofing going up. Perez Mason's preaching stand had been removed and the benches, dating from 1846 and 1851, were stored, to be put back after completion. Three Sunday services were held in

The Iron Tabernacle, interior. (Alison Shaw)

the new tabernacle July 28. The day was beautiful, following rains, and nature was "vocal with praise."

The ingenuity and appeal of Wesleyan Grove's iron tabernacle begs for clues about the background and aesthetic intentions of its designer. The firm of Dwight and Hoyt was a brief partnership between George C. Dwight, a builder and dealer in corrugated iron who does not figure directly in this story, and his employee, John William Hoyt (1839–1911). A civil engineer and campground resident, Hoyt was born at Sag Harbor, Long Island, the son of a Methodist minister. He graduated from Wesleyan University in 1858, presumably in the new scientific course, and by 1871 was established in Springfield.[30] He must have been well connected to inner camp-meeting circles, for he was a distant relative of the famed—even notorious—Kate Chase Sprague, daughter of Chief Justice Salmon P. Chase, Lincoln's Secretary of the Treasury, and wife of U. S. Senator William Sprague, Rhode Island textile magnate, who owned one of the campground's largest cottages. Sprague's personal involvement with Wesleyan Grove seems minimal, limited to one year. He did not need a summer home, having built a vast waterside place in Rhode Island, but he may have been attracted to Methodism for its reputation for controlling alcoholism, a Sprague problem. John W. Hoyt was also cousin of rich, high-ranking New York Methodists who often visited Wesleyan Grove from their yacht. Whatever the inner connections, Hoyt got to build an elegant but nonconforming cottage on valuable lots on the main circle, near the Sprague house. Hoyt's graceful, sophisticated house is both Pompeiian, with a central atrium room with raised roof and encircling clerestory, and Italian with its two-story tower with flattened hip roof and bracketed eaves.

Just as Hoyt's bare biography does not account for his talent, it is also hard to place the tabernacle in its rightful niche in the history of American metal building construction. The most obvious precedents, such as railroad stations, were much larger. The building is probably best understood as a translation into metal of a wooden building type, the hipped-roof "arbor" of southern camp meetings. Tabernacle drawings by E. N. Boyden, one of the three competitors, survive and show a structure between Hoyt's design and the simpler southern types. Boyden proposed three tiers of roof resting on braced wooden supports, the top roof octagonal, the lower two square in plan, reverse of the iron tabernacle. Both designs have top clerestories which are taller than the lower ones, and both have arched, gabled entrances set into the lowest roof. The two surviving Boyden projects show simple and more elaborate versions of the same building, the second being overtly Gothic with quatrefoil openings in the cupola drum and little flying buttresses connecting it to the roof. The simpler project, without cupola and buttresses, is still more deliberately detailed than any southern tabernacle. With this drawing or one similar as the tabernacle's parent, we probably only need to know about the Phoenix Iron Company's production in 1879 to understand whether standard or specially made components were used and how Hoyt devised it.

Wesleyan Grove's iron tabernacle was a critical success among the limited number of people who wrote about it. Built after the wave of ecstatic campground accounts of the 1860s and early 1870s, it was little known off the island even in modern times. It was not reported in the professional press until 1973. Photographs could not capture its grand spatial beauty the way they could record the aggressively detailed planarity of the cottages and the sculptural prowess of S. F. Pratt. At dedication it was called a "beautiful building," not at all "barn-like," and its acoustics were admired. One writer saw it as "a kind of religious crystal palace, though made of iron." Another made the more subtle observation that there was no ornament, "the whole effect of the building being produced by the lines of construction."[31] But no debate ensued over whether its beauty or lack thereof indicated a greater or lesser religious moment, whether it spoke of social growth or decay. Like so many of the world's monuments, the cultural energy that went into its creation had already crested, leaving the structure as the memento of collective spirit right at the beginning of its long life as a successful, working building.

The end of a phenomenon is always harder to study than its beginning, for no one cares to document it. By the 1880s Wesleyan Grove and Oak Bluffs had long ceased to be the focus of transient tourism and journalistic amazement. The Oak Bluffs column in the *Providence Daily Journal* had now been moved from the front page to join other suburban notes at the back of the paper. In the early 1880s the revival subsided into the "tabernacle meeting," week-long sessions of sermon and prayer and a closing Sacrament. We know something about the 1885 meeting because it was Wesleyan Grove's semicentennial. Grace Chapel, the last association building, was dedicated for use by the women. The founders of the camp meeting were praised and the conversion of a Jew remembered. Twenty-one people rose to claim their presence in 1835, and two who were converted at that distant time were honored. Bishop Foster, now the owner of Bishop Haven's elegant cottage on Clinton Avenue, admired the grounds, saying that no other place on the continent had such a tabernacle and park. One observer said the association was a pioneer of rural improvement societies. But Bishop Foster also spoke against revivals. Tabernacle services, he said, elevate and broaden, but revivals are a matter of noisy services and religious crudities. Even so, "The preacher closed in a style which set the colored

Fireworks in Ocean Park, 1985. (Mark Lovewell)

people shouting, and not a few of the whites joined them." The *Journal*'s correspondent thought it a genuine camp meeting.

By 1885, the town of Cottage City, later Oak Bluffs, had achieved the physical and social form it would maintain for decades. Families had become committed to their cottages and summer way of life, and their traditions would continue with only some loss until the tourist boom of the present day. "Oak Bluffs always gives you the impression of a place where a great many people are having a tremendously good time together" it was written in the 1920s. By then the "raw newness" of the resort was softened by a covering of trees, making it more like the campground, as "embowered as a college campus."[32] Trees compensated for the loss of the involving but prickly Victorian detail to twentieth-century shingling. The twentieth century saw the expansion of the year-round Azorean Portuguese farming and fishing population and the arrival of summering well-to-do blacks from Boston and Harlem.[33] They would keep the town vital and contribute to its relaxed spirit. But that is another story. For 1885, we must be content to read of the mourning for President Grant and the Hon. Oliver Hoyt's melon-and-cream party at the Clinton Avenue cottage of the Reverend Dr. Tiffany. Oak Bluffs, encapsulating Wesleyan Grove and its revivals, had settled into a long and happy afterlife.

SEVEN

CITY IN THE WOODS

THE STORY OF WESLEYAN GROVE was of the rapid growth of a temporary religious community from a small circle of tents in 1835 to the canvas "city" in the woods of 25 years later, a place of fame and great physical charm, locus of extraordinary experience, and the object of deep affection of generations. By 1860, a transformation was under way, the "celestial city" of tents in the woods, like "flocks of white birds resting in the shade," becoming a city of cottages of comparative material splendor and remarkable visual quality. The cottages, invented at this site, were placed close together along the narrow lanes and around little parks to make a "mazy" effect of "sweet disorder," which was astonishing to the increasing numbers of visitors who came from ever greater distances to see as well as to worship. Visitors found themselves transported into an unimaginable world, a "fairy-land," but one still dominated by nature and thus able to function for its original revival purposes. Wesleyan Grove became a permanent summer community which was still a religious space, and it achieved this without a professional designer. It was a popular or folk created work of environmental art.

It is with Oak Bluffs, the resort developed next to the revival at the end of the 1860s, that designers—a landscape architect and an architect—were enlisted to help shape the place into something imageable and excellent. Design skills applied lessons from Wesleyan Grove to unite the resort formally and thus spiritually to the old campground, helping maintain relations between the religious and secular institutions. Design skills built upon the exotic otherness of the campground to extend the sense that this island place was magical, a figment of the imagination, without equivalence in human experience. Campground and resort were retreats from city life into nature, but also retreats from farming and seafaring isolation into society. With their festive human pageantry and architectural fantasies of turrets, towers, pinnacles, and wooden lace, they were pure and beloved expressions of an ideal urbanity. The poles of city versus country were obliterated and a dream made real by the crowds under the trees, and before the sea.

Considering the visual power of the architectural forms at Wesleyan Grove and that the community was so well known by so many, it is surprising that the place was rarely described in the nineteenth century using the language of criticism. But aesthetic terminology is usually restricted to intellectuals, for whom so much of what the meeting represented was repugnant, as we have seen with Nathaniel Shaler and Andrew Dickson White. James Jackson Jarves could well have used art criticism to deal with it but did not, perceiving it rather as an experience of which art was only one component. An anonymous writer in the *Camp Meeting Herald* in 1866 wished that he had the pencil of Turner and the pen of Byron or Ruskin to describe the mixture of man, nature, and joy of the Grove.[1] But this allusion to Romantic masters is about the only example at hand. The campground resident of the greatest literary ability, Bishop Gilbert Haven, was an impassioned admirer of Ruskin, whom he viewed as the greatest Christian writer since Milton and, because Ruskin dealt with nature as scientist and as poet, the consummate nineteenth-century man.[2] But Bishop Haven did not write about Wesleyan Grove, at least as far as we know. In 1882, there is one fine but unattributable description in which the new stylistic term of the "Queen Anne" and considerable architectural sophistication is applied to Oak Bluffs. It is worth quoting at length for its evocation of an aesthetic and because it speaks in almost uncanny fashion to our postmodernist and preservationist sensibilities. There will be, we read, those in a distant future who, having forgotten the nineteenth century and tired of their own ascetic times, will yearn for more.

> And what an interesting pursuit, indeed, their researches will prove. In these Arcadian days that are to be how gladly those eyes fatigued with long searching after the architectural "Unities" will welcome the new light com-

> ing, as it does by the many angled reflections of the past, from a star that has already set; with what joy they will learn that in the good old days of Queen Martha, the most noted city of her Island had regard, in its dwellings, to nothing more imperative than whimsical caprice; that the sternest exactions known to the prevailing style were hospitality, comfort, and abandon.
>
> Then how vain and full of shortcomings will be the imitation attempted by posterity. . . . Taken altogether, their cities will not be like our city, nor their dwellings like our dwellings.[3]

These are nineteenth-century anticipations of the pleasures the twentieth century finds in Victorian building, but stated in a way that twentieth-century "Victorians" do not quite dare to do.

A late twentieth-century art historical view of the situation is slightly different. Wesleyan Grove appears as essentially romantic, with its exoticism, nature immersions, and appeals to feelings. So was Methodism, the religion of the heart. The medieval styling of the cottages is, too, part of romanticism; but one is also struck by the classical formal elements just below the stylistic surface. The planarity of the cottage facade is its dominant visual quality. The vertical striations and sharp contrast between the wall plane and the hood molds or jigsaw trim represent a linearity that is in keeping with the building type's folk origin. The cottages are a contribution to the wish of art historian Barbara Novak to find classicism as the most continuous thread of American art. A classical underpinning for a romantic situation corresponds nicely to the recurring descriptions of the people as quiet, staid, and restrained, even in their expressions of joy.

Oak Bluffs, if one takes the wide expanse of park and sea as the counterbalance to all the woodworking display, is also part of the same fabric of feeling as the flat-faced cottages. Ocean Park, ostensibly a real estate developers' maneuver to ensure lot sales, made an air-and-water panorama of hypnotic grandeur. Robert Rosenblum, describing nineteenth-century American landscape painting (and twentieth-century abstraction), was caught by the "strange, haunting openess," the "uncanny quiet," the "primordial void of light," and the "empty unlocalized space" that characterized our artistic vision.[4] These phrases can apply as well to Oak Bluffs's vista of park and sea—empty, luminous, awesome, even fearsome. Just as American art never gave itself completely to the painterly surfaces of French Impressionism, the American environment remembered the silence and the void of space and light.

Another issue to be illuminated by Wesleyan Grove and Oak Bluffs is that of Victorian wealth and material display. Late eighteenth-century Methodists were a plain people, and asceticism and attention to the poor were critical values. The *Discipline* pronounced it impossible for a rich man to enter heaven and urged that chapels be simple and inexpensive so that Wesleyans could avoid depending on the wealthy. New England, however, was always lax with codes of austerity. Bishop Asbury and Jesse Lee were dismayed by the steeple and bell on a Newport meetinghouse, and pew rental, proscribed by the General Conference because it put the poor in back rows, was practiced in Boston at the beginning of the century. When frontiersman Peter Cartwright was in Boston in 1852 for the General Conference that ended the old free seat rule, he disapproved of that decision along with the "old wooden god the organ, bellowing up in the gallery, and the few dandified singers" who seemed to be replacing vigorous congregational hymn singing.[5] By the 1850s Methodists all over were getting rich. In 1855, an article on Methodist city church architecture rejected the old strictures on ambitious buildings, arguing that a meetinghouse should not look like a barn, school, or house. This piece provided plans and elevations for two big churches, one Romanesque and one Gothic, both with elaborate towers, spires, and finials. The author urged the anti-steeple faction among his readers to "sheath their swords" and read on in good spirits, for churches should correspond to the houses of the worshipers and these were becoming very fine.[6] New Bedford journalists SAM and Kirwin's debate about the neglected society tents and fancy cottages was part of a larger issue. Finally, in 1872, the *Discipline* was changed to eliminate the barriers to heaven for the wealthy.

Many of the residents of Wesleyan Grove surely must have felt themselves part of this half-century ascension from poverty to wealth. While more Oak Bluffs men than campground residents acquired published biographies, they probably represent a fair sample of those involved. Albert S. Barnes, the Brooklyn publisher, had a Spartan childhood on a New England farm, apprenticing to a printer in Hartford, where he boarded at a pious Christian home. By 1880 he owned his large island cottage and a big house in Brooklyn, and was active in Sunday school education for the urban poor. Philip Corbin, one of many children of a straitened farmer, was apprenticed to a locksmith and eventually co-owned the P. and F. Corbin Company, one of the largest manufacturers of architectural hardware of the nineteenth and the twentieth centuries. George Landers, son of a teacher, started out as a carpenter but became a manufacturer of durable household goods and a state politician. Isaac Rich went from orphan fishmonger to millionaire. The biographies emphasize their limited common school education, ceaseless industry, and religiously based habits of mind which propelled them to success and cushioned them in failure. Tim Stanley, a New Britain plated-goods manufacturer, loved nature, art, and his Oak Bluffs home. He weathered business reverses that would have felled any other man because of his philosophical character. Pardon M. Stone, a campground resident and member of the finance committee, went through many terrible losses and yet, because of his farming boyhood close to nature and his religion, remained a remarkably unembittered man. It is against this background of little education, rural boyhood poverty, and acquired wealth that we must see the assertive building forms and even the urge to cluster together in a semiurban community for summer release. In doing this they made a collective monument, the repository of intense feelings and cherished values, which softened the anxiety of geographic mobility or economic failure. Architectural historian Walter L. Creese has recently shown that American artist-intellectuals from Jefferson to Downing, Olmsted, and Wright wanted to give raw America security and psychological comfort as well as vision and nobility. The dynamism of American life demanded solace as well as celebration. The residents of Oak Bluffs and Wesleyan Grove, armed with a lesser literature and middling rather than great artists, through collective effort, achieved both.

Nineteenth-century America was a vast raw continent in which people were uncommonly busy at the art and enterprise of making communities. Europeans did not know this task, having inherited their place, unless, of course, they left to join the overseas drama. New American communities were of a startling variety—gold rush boom towns, ambitious plains "cities" scrabbling for the certainty of county seat or state capitol, self-contained religious communes, self-conscious Utopian experiments. By the middle of the century, Americans were planting summer resorts, semiprivate residential enclaves, industrial villages, rural colleges, camp meetings, and rural cemeteries in quantity, all showing a mix of hard-edged business acumen powered, or at least softened, by the kind of idealistically shaded goals that we know from Wesleyan Grove and Oak Bluffs. Only now are other highly individual communities which show an undercurrent of religious idealism being uncovered. Vineland, New Jersey, for example, was a temperance agricultural colony established for immigrants trapped in the cities.[7] It also hosted the first meeting of the reformist National Camp Meeting Association, a Holiness league dedicated to the primitive camp meeting goal of Wesleyan perfection or the second blessing. Swedenborgian-/Transcendental Llewellyn Park, New Jersey, and Presbyterian Lake Forest, Illinois, were midcentury proto-suburbs sharing design attitudes with Wesleyan Grove. A similar creative tension must have existed in these communities—between materialism and spiritual fervor, human density and nature, community and privacy, and intentional design and accident. Camp meeting, resort, and residential enclave are intertwined in their parentage of the twentieth-century suburb, the final celebration and solace for the family, if not always the community.

America's preferred residential environment, the single-family house among others in nature, has been given short shrift in the literature, in part because of the Euro-centric bias of so many of our design critics and educators, both European and American. These travelers know too well how the charms of European residential cities, with their pedestrian vitality and plazas and cafés, can compensate for the fact that nature, well-tamed, is restricted to parks, allotment gardens, and the well-used countryside. Their views have been bolstered by American intellectuals, whose annoyance with the middle class is so consistent as to suggest the need for a study that might be called "The Intellectual Against the Suburb," a modern sequel to the Morton and Lucia White classic, *The Intellectual Versus the City.* Indeed, the work has been well started. In an unpublished dissertation of 1971, Robert Winston Taylor surveyed the antisuburb literature of the middle twentieth century, from Lewis Mumford in 1921 through the crescendo of the 1950s and 1960s in the writing of John Keats, John Seeley, David Riesman, C. Wright Mills, William White, Peter Blake, A. C. Spectorsky, and others.[8] A beginning list of the evils of civilization spawned by suburban life as culled from Taylor's collection, arranged alphabetically, follows: adultery, alcoholism, alienation, anxiety, apathy, breakdown of community, compulsive consumerism, conformity, credit buying, destruction of family, destruction of individual morality, divorce, facelessness. . . . The fierceness of the rhetoric precludes any attempt at sympathetic understanding, but does highlight a broad stream of antisuburban bias endemic to modern American thought. Taylor rather mildly suggests that perhaps the middle-class American intellectuals of the period after World War II were feeling displaced by the upwardly mobile working class then establishing itself in vast new housing tracts.

Stable, older, tree-crowned suburbs are America's finest common environmental creation. They look good not only because of the contrast to trashy highway strips, the paved voids of shopping mall parking lots, and decimated and scaleless city centers, but also all on their own. They rank with such residential glories as Italian hill towns, the boulevards of Paris, and English terrace-house squares and crescents. Our appreciation of suburbs as environmental art is new and immature. We have yet to comprehend their human and aesthetic systems. Observations from Wesleyan Grove and Oak Bluffs may help. Our suburbs are intended as social environments, although some critics may feel that they fail at the task. They reflect not only the urge to flee the city into "places of nestling green," but also the communal values of social cohesion and interdependence, man-

Cottage in the campground. (Ellen Weiss)

ifestations of earlier Utopian thought.[9] The pre-Civil War urge to collectivism, whether theoretical in Utopian forays or folkish and biblical in camp meetings, was made permanent by the pragmatic late nineteenth century middle class, only to be transformed again by their progeny, the first to weaken the communal dimension. Yet the modern leafy descendants retain much of the nineteenth-century idealism. Although each house on its lot is held by the individual family in its own natural compound, the view from house to house, and symbolically from family to family, may be as important as the opportunity to garden or to store a boat or simply to own land. Reaffirmation of the self through watching others who share one's values is a necessary human task. The blatancy of this activity at Wesleyan Grove and Oak Bluffs should help us understand.

Another lesson from Wesleyan Grove and the forest revival is the psychic benefit of nature immersions. Greenery, gardens, dappled shade, and all the picturesque irregularities take the strains of the day from the tired commuter. When stormy winds tangle the suburban tree branches overhead and blow leaves into untended corners of the lawn, a breath of the wilderness returns to the middle landscape. We can feel for a moment the release the campers at Methodist revivals must have known. Suburbs provide a range of nature experience, from well-manicured gardens replete with emblems of personal taste in ornaments and plantings to the easy mess that happens so quickly when the householder's energies are directed elsewhere. A house buried within greenery is balm for the spirit, and intimately bound in with being American. One cannot imagine the Transcendentalists any place but Concord—village outside of city—gardening between bouts of hard intellectual work, putting aside books for nature because nature is available in hourly alternation with focused activity. The less intellectual residents of Wesleyan Grove did their hard work in groups at the preachers' stand, but the relief of unpredictable patterns of wind-rustled leaves, and their shadows on the sand, just outside the cottage door, was always there. Crowds of people, citylike crowds, all immersed together in the green, was always the astonishment of Wesleyan Grove. It was the intention in the suburb that something of this be made to live on.

NOTES

ABBREVIATIONS USED IN NOTES

DCHS	Dukes County Historical Society
Gazette	*The Vineyard Gazette*
Hough	Henry Beetle Hough, *Martha's Vineyard, Summer Resort.* Rutland, Vt.: The Tuttle Publishing Co., 1936
Journal	*The Providence Daily Journal*
JSAH	*Journal of the Society of Architectural Historians*
Mercury	*New Bedford Mercury*
Standard	*New Bedford Evening Standard*
Vincent (1858)	Hebron Vincent, *History of the Wesleyan Grove Camp Meeting from the First Meeting Held There in 1835 to That of 1858.* Boston: George C. Rand and Avery, 1858
Vincent (1870)	Hebron Vincent, *A History of the Camp Meeting and Grounds at Wesleyan Grove, Martha's Vineyard, for the Eleven Years Ending With the Meeting of 1869. . . .* Boston: Lee, 1870

CHAPTER ONE

1. Charles M. Johnson, *The Frontier Camp Meeting* (Dallas: Southern Methodist University Press, 1955), is a detailed and provocative study of all aspects of the western camp meeting during the first third of the nineteenth century. The present study is indebted to this and to the brief discussions of camp meetings in the following general works: William Warren Sweet, *The Story of Religions in America* (New York: Harper and Bros., 1930); William Warren Sweet, *Methodism in American History* (New York: The Methodist Book Concern, 1933); Wallace Guy Smeltzer, *Methodism in the Headwaters of the Ohio* (Nashville: The Parthenon Press, 1951); and Sydney E. Ahlstrom, *A Religious History of the American People* (New Haven: Yale Univer-

sity Press, 1972). Also important have been Dickson D. Bruce, Jr., *And They All Sang Hallelujah* (Knoxville: The University of Tennessee Press, 1974); and Bernard Weisberger, *They Gathered at the River* (Boston: Little, Brown, 1958). A clear and succinct history of Methodist doctrinal issues is to be found in *The Book of Discipline of the United Methodist Church* (Nashville: The United Methodist Publishing House, 1976). Also useful for a general understanding of Methodism in the nineteenth century are Richard M. Cameron, *Methodism and Society in Historical Perspective* (New York: Abingdon Press, 1961); Timothy L. Smith, *Revivalism and Social Reform* (Nashville: Abingdon Press, 1957); William G. McLoughlin, *Modern Revivalism* (New York: Ronald Press, 1959); and his *The American Evangelicals* (New York: Harper and Row, 1968).

2. Smelzer, p. 105, and Johnson, p. 85.

3. For Cane Ridge see Johnson, pp. 61–67, and Sweet, pp. 156–158. For the English rejection of camp meetings, see Johnson, p. 120 and, in addition, Richard Carwardine, *Transatlantic Revivalism* (London: Greenwood Press, 1978) and Elisabeth K. Nottingham, *Methodism and the Frontier, Indiana Proving Ground* (New York: Columbia University Press, 1941).

4. For an account of the process of experiencing Holiness, see the Reverend B. W. Gorham, *Camp Meeting Manual* (Boston: H. V. Degen, 1854), pp. 85–87.

5. *Western Christian Advocate,* I(16), 15 August 1834.

6. Nathan Bangs, *A History of the Methodist Episcopal Church* (New York: T. Mason and G. Lane, 1838 and 1839), II, pp. 113, 116, 118.

7. Bangs, pp. 271, 272.

8. Bangs, pp. 274, 275.

9. Gorham, pp. 30, 32.

10. *Western Christian Advocate,* VI(271), 5 July 1839.

11. Gorham, p. 38.

12. *Standard,* 18 August 1861, and *Mercury,* 20 August 1861.

13. Gorham, pp. 31–32.

14. Jesse Lee, *A Short History of the Methodists in the United States of America* (Baltimore: Magill and Clime, 1810), pp. 360–362.

15. Bangs, II, p. 266.

16. John F. Wright, *Sketches of the Life and Labors of James Quinn* (Cincinnati: Methodist Book Concern, 1851), p. 120. *Western Christian Advocate,* IV(14), 28 July 1839, and I(16), 15 August 1834.

17. Talbot F. Hamlin, *Benjamin Henry Latrobe* (New York: Oxford University Press, 1955), pp. 319–320.

18. The Campground Cemetery near Farmer City, Illinois, is an example. For another campground cemetery see John B. McFerrin, *History of Methodism in Tennessee* (Nashville: Southern Methodist Publishing House, 1871), II, pp. 310–311. McFerrin also mentions campgrounds acquiring permanent churches.

19. John Brinckerhoff Jackson, "The Sacred Grove in America," *The Necessity for Ruins* (Amherst: The University of Massachusetts Press, 1980).

20. Frances Trollope, *Domestic Manners of the Americans* (London: 1832), pp. 240–245. Reprinted in Johnson, Appendix I, pp. 255–257.

21. Wright, p. 108.

22. Thomas L. Nichols, *Forty Years of American Life* (London: John Maxwell and Co., 1864), I, p. 80.

23. Bangs, II, p. 265.

24. "A genius like his found especial delight in the camp-meeting. Its freedom from

restraint, its communion with Nature, the exhilaration of opposition, its largeness of life, where every noble impulse is itself ennobled, all combined to make him an ardent lover of its services. Almost his first pulpit triumphs were on this field; and, to his last days, he cherished a warm attachment for its altars." The Reverend Gilbert Haven, *Father Taylor, the Sailor Preacher* (New York: Phillips and Hunt, 1871), p. 222.

25. George Prentice, *The Life of Gilbert Haven* (New York: Phillips and Hunt, 1883), p. 180.

26. Johnson, pp. 245, 246.

27. *Gleason's Pictorial Drawing Room Companion,* IV(11), 11 September, 1852, p. 176.

28. Information on camp meetings in Virginia, South Carolina, Tennessee, Kentucky, and Louisiana has been kindly furnished by state historic preservation offices in the form of nominations to the National Register of Historic Places. North Carolina information has been furnished by the Historic American Buildings Survey and by Bernard Herman. Kenneth Thomas of the Georgia Department of Natural Resources led the author to Shingleroof and Salem. The Mississippi meeting, Spring Hill, was reported by Robert M. Ford in *Mississippi Houses: Yesterday Toward Tomorrow* (privately printed, 1982).

29. Peter Cartwright, *Autobiography of Peter Cartwright, The Backwoods Preacher,* (Cincinnati: Cranston and Curtis, n.d.), p. 45; Johnson, p. 48.

30. William M. Clements, "The Physical Layout of the Methodist Camp Meeting," *Pioneer America,* 5(1), January 1973, pp. 9–15, attempts to show that two North Carolina meetings, Tucker's Grove and Rock Springs, as presented in National Register nominations, illustrated Gorham's *Manual.* These meetings are older than the book and have the wooden cabins and large tabernacles typical of the South, rather than the tents and preaching stand of the North.

31. Gorham's print can be found in James Mudge, *History of the New England Conference of the Methodist Episcopal Church* (Boston: The Conference, 1910), plate VI; Dickson D. Bruce, Jr., *And They All Sang Hallelujah,* plate XVI; and Johnson, opposite p. 114.

32. Gorham, p. 137.

33. Gorham, pp. 163, 164–165, 166, 168.

34. For the history of Methodism in New England, see Jesse Lee, *History of Methodism* (Baltimore: Magill and Clime, 1810); Mudge; and Rennetts C. Miller, *Souvenir History of the New England Southern Conference* (Nantasket: The Conference, 1897).

35. Bangs, II, p. 7.

36. Stephen Allen, *The Life of Rev. John Allen, Better Known as "Camp Meeting John"* (Boston: B. B. Russell, 1888).

37. Jeremiah Pease, "The Island's First Methodists" [1847], *The Dukes County Intelligencer,* 22(2), November 1980, pp. 58–70.

38. Pease, pp. 62, 63–67.

39. John Adams, *The Life of "Reformation" John Adams* (Boston: George C. Rand, 1853), I. Chapters XII and XIII are an account of Adams's first years (1821–1822) on the island. In 1826, again on the Vineyard, he endured a nervous breakdown in which he had visions of southern New England as conflicting biblical geographies. The Vineyard became Patmos with sections of it Goshen, the plains of Moab, or Canaan. Jeremiah Pease became Shadrach. For a fine understanding of what it meant to be a Methodist in Congregational Edgartown, see Arthur R. Railton, "Jeremiah's Travail," *The Dukes County Intelligencer,* 22(2), November 1980, pp. 43–57.

40. Jeremiah Pease's diary is being published in segments in *The Dukes County Intelligencer,* beginning November 1974, 16(2).

41. Autobiographical material by Hebron Vincent (1805–1890) is contained in a lengthy

letter written upon election to the New England Methodist Historical Society, 11 October 1880, a copy of which is filed in Box 87B, DCHS. Additional notes on his life are in *Alumni Record of Wesleyan University,* 3d ed. (Hartford: Case, Lockwood, and Brainaird, 1883). Vincent never achieved his goal of attending the Middletown seminary because of poverty and illness, but he received an honorary Master of Arts degree in 1869. Illness also cut short his preaching career in the New England and Providence conferences. He published extensively on horticulture, education, and Methodism, and was Register of Probate in Edgartown, a member of the Massachusetts Board of Agriculture, and represented clients in claims in Mexico.

42. [Samuel Adams Devens], *Sketches of Martha's Vineyard and Other Reminiscences of Travel at Home, Etc.* (Boston: James Munroe and Co., 1838).

43. Reverend John B. Gould, "The Great Martha's Vineyard Revival of 1853," in Miller, XXIV–XL.

44. Johnson, p. 211.

CHAPTER TWO

1. Vincent (1858), p. 5.

2. Vincent (1858), pp. 20, 23–24; Ninety-five-year-old Sarah Pease, interviewed in 1923, believed that there were 65 attending in 1835 (*Standard,* 22 July 1923).

3. Hough, p. 9.

4. Vincent (1858), p. 113.

5. Vincent (1858), pp. 65, 82.

6. Vincent (1858), pp. 80, 123–124.

7. Vincent (1858), pp. 49–50; James Mudge, *History of the New England Conference of the Methodist Episcopal Church* (Boston: The Conference, 1910), pp. 392–393.

8. Sirson P. Coffin, *Annual Report of the Agent of the Martha's Vineyard Camp Meeting* (New Bedford: privately printed, 1860), describes the meeting at this time in terms of financial poverty. It took several years for the Edgartown sponsors to pay off their debts on the new arrangements. The only known copy of this document is in the New Bedford Free Library.

9. Vincent (1858), p. 169.

10. Vincent (1858), p. 191.

11. Vincent (1870), pp. 31–32. Notes by Henry Beetle Hough in the *Providence Sunday Journal,* 5 August 1945, p. 4., and *Gazette,* 3 June 1870.

12. For a radial concentric plan by 1860 see *Mercury,* 21 and 24 August, 1860, and Vincent (1870), pp. 32–33.

13. *The New York Times,* 20 August 1867; and *The Tourists' Guide to Southern Massachusetts* (New Bedford: Taber Brothers, 1868).

14. Wallace Guy Smeltzer, *Methodism in the Headwaters of the Ohio,* p. 108, cites an early campground of "many circles of tents, divided by narrow streets and alleys." For Sirson P. Coffin's involvement with Round Lake see the *Journal,* 16 August 1865. Round Lake histories generally credit the Vineyard meeting as inspiration for its foundation. See Arthur James Weise, *History of Round Lake* (Troy: Douglas Taylor, 1887) and George Hughes, *Days of Power in the Forest Temple* (Boston, 1873), p. 59. For Pitman Grove, Charles Stansfield, "Pitman Grove: A Camp Meeting as Urban Nucleus," *Pioneer America,* VII(1), January 1975. A tradition of allusion to Revelation is in the American Guide Series, *New Jersey* (New York: Viking Press, 1939), p. 658. Other views of this interesting community can be found in Harold F.

Wilson, *Cottages and Commuters, A History of Pitman, New Jersey* (Pitman: Pitman Borough, 1955); the 1975 National Register nomination; and a 1963 Historic American Building Survey report. For Lancaster, Ohio, and Plainville, Connecticut, see the National Register nominations. For Washington Grove, see the 1978 National Register nomination. For Des Plaines, Illinois, see Almer M. Pennewell, *The Methodist Movement in Northern Illinois* (*The Sycamore Tribune*), 1942, pp. 273–274. The first meeting at Des Plaines in 1860 had a "spokes-in-a-wheel" plan by master builder and architect James Lawrence. The meeting moved to its present site in 1865, but repeated the wheel plan.

15. John R. Stilgoe, *Common Landscape of America, 1580–1845* (New Haven: Yale University Press, 1982), pp. ix, 9. There is an excellent chapter on camp meetings in this volume. The Gourlay plan is reproduced and discussed by Walter Muir Whitehill in *Boston, A Topographical History* (Cambridge: Harvard University Press, 1968), pp. 146–147. It is also published in John W. Reps, *The Making of Urban America* (Princeton: Princeton University Press, 1965), p. 292, but is incorrectly identified as Robert Morris Copeland's 1872 Boston plan.

16. *Journal,* 21 August 1863; Vincent (1870), p. 34; *Mercury,* 7 August 1862, *Camp Meeting Herald,* 6 August 1862.

17. *Camp Meeting Herald,* 9 August 1862.

18. Vincent (1870), pp. 41–43.

19. Perez Mason obituary is in *Cottage City Star,* 17 March 1881. For the stand and seating plan see *Journal,* 29 August 1860 and 17 August 1861; *Standard,* 13 August 1861; and Vincent (1870), pp. 63–64.

20. Vincent (1870), p. 115, and *Gazette,* 24 July 1864.

21. *Mercury,* 23 August 1864.

22. *Seaside Gazette,* 1 September 1873; *New York Herald Tribune,* 26 August 1868; Rev. Gilbert Haven, quoted in Hough, p. 130.

CHAPTER THREE

1. The 1860 figure of 500 is from a bird's-eye view drawn in 1878–1879 and published in 1880 by Sunderland of Providence. The figure needs verification in camp-meeting records. By 1914, cottages were on the decline because many were moved to other locations. A Sanborn insurance map for that year shows about 364 cottages. There are 306 listed on the 1978 National Register nomination form.

2. For an introduction to the Gothic Revival in America see Leland M. Roth, *A Concise History of American Architecture* (New York: Harper and Row, 1980), and William H. Pierson, Jr. *American Buildings and Their Architects: Technology and the Picturesque, the Corporate and Early Gothic Styles* (Garden City: Doubleday, 1978).

3. An illustrated survey of southern New England Methodist churches is provided in Rennetts C. Miller, *Souvenir History of the New England Southern Conference* (Nantasket, Mass.: 1897). The article on "Methodist Church Architecture" is in *The National Magazine,* December 1855, pp. 497–512. For an account of the history if not meaning of the mid-nineteenth-century Romanesque in America see Carroll L. V. Meeks, "Romanesque Before Richardson in the United States," *Art Bulletin,* 35, March 1953, pp. 17–33.

4. *Gazette,* 28 May 1869.

5. Ernest Allen Connally has furnished the clearest description of "plank wall construction" in "The Cape Cod House: An Introductory Study," *JSAH,* 19(2), May 1960, pp. 47–56. Also valuable is the analysis of "vertical board plank frame construction" in seventeenth-

century Plymouth Colony by Richard M. Candee in "A Documentary History of Plymouth Colony Architecture, 1620–1700: Construction Methods," *Old-Time New England,* 60(2), October–December 1969, pp. 36–53. For an account of a similar structural system in the nineteenth century see Diane Tebbetts, "Traditional Houses of Independence County, Arkansas," *Pioneer America,* 10(1), June 1978, pp. 36–55. For an eighteenth-century Vineyard house with studs see Anne W. Baker, "The Vincent House: Architecture and Restoration," *The Dukes County Intelligencer,* 20(1), August 1978, pp. 7–27.

6. Because the "cottage" and "tent" distinctions were made in the campground records for the years back to 1864, the association has believed that the first cottage was built in 1864 and has put a bronze plaque on the Mason-Lawton cottage with the date 1864. All other nineteenth-century sources indicate an 1859–1860 date for the building.

7. The first mention of the Reverend Frederick Upham's "rough boarded tent" is in the *Standard,* 17 August 1860. Hebron Vincent, writing of the cottage developments of 1866 (Vincent [1870], p. 145) says that the first wooden building was Upham's, and that it was built "ten to twelve years previous" and was 10 by 12 feet. This would put the building at 1854 to 1856. One island scholar is now arguing for a date of 1851. (See editorial note by Arthur R. Railton, *The Dukes County Intelligencer,* May 1985, p. 184.) *The Camp Meeting Herald* also named the Upham cottage as the "first family residence entirely of wood" (August 17, 1869). A postcard of the building, with its vertical planks exposed, is in the collection of the late Stuart Sherman, Providence, RI. The building is now part of the kitchen at 12 Cottage Park. Frederick Upham, born 1799, was a shoemaker who started preaching in 1820 and kept at it for 70 years. His son, Samuel Foster Upham, who shared his Vineyard cottage, was educated at Wesleyan University but was still characterized as a fervent Methodist preacher of the old-time sort, a "comrade of the heart." He was a pastor for 24 years and then taught at Drew Seminary for another 24. Both men were well known all through the New England conference.

8. *Camp Meeting Herald,* 17 August 1869.

9. *Standard,* 17 August 1860. Another early reference to this cottage is in the *Journal,* 17 August 1861. Vincent included it in his account of the year 1860 (Vincent [1870], p. 49).

10. For an introduction to A. J. Downing and his importance in American domestic architecture see David P. Handlin, *The American Home* (Boston: Little, Brown, 1979). p. 34–46.

11. The Aaron Siskind collection is at the Dukes County Historical Society. For the Cook cottage, see [Hebron Vincent] *The Vineyard as it Was, Is, and Is To Be, By an Observer* (New Bedford: privately printed, 1872), p. 41. Another source (*Journal,* 13 August 1863) has it built in 1863.

12. *Standard,* 4 August 1862; *Mercury,* 4 August 1862.

13. *Mercury,* 13 August 1862.

14. *Camp Meeting Herald,* 22 August 1866; *Standard,* 18 August 1866; David Clark, *The Way Reverend Moses L. Scudder Secured a Cottage at Martha's Vineyard* (Hartford: privately printed, 1870), p. 11.

15. Clark, p. 7.

16. *The New York Times,* 26 August 1866, 20 August 1867.

17. *The New York Times,* 20 August 1866.

18. *Standard,* 18 August 1866 and 23 July and 15 August 1867; *Journal,* 31 July 1869.

19. *Gazette,* 25 September 1868; Stuart D. Ludlum, *Exploring Nantucket, Martha's Vineyard 100 Years Ago* (Utica: Brodock and Ludlum Publications, 1973), p. 21; *Cottage City Star,* 8 July 1882; *Standard,* 13 August 1864.

20. N. Robinson, "The Islands of the Bay State," *Frank Leslie's Popular Monthly,* 12(1), July 1881, pp. 103, 102.

21. There are large collections of stereoptican cards at the DCHS and at the Society for Preservation of New England Antiquities, Boston, A remarkable find of about 1,000 glass negatives taken by Woodward and Son of Taunton, Mass., in the 1870s and 1880s is owned by Vineyard Vignettes.

22. *Standard,* 18 August 1866 and 25 August 1870; *Journal,* 18 August 1871; *Gazette,* 5 August 1961.

23. *Gazette,* 2 August 1867 and 18 August 1976; Vincent (1870) p. 325.

24. Vincent (1870), p. 145 Biographical material on island carpenters comes from the Roy Norton Genealogical Collection at DCHS, the town records of Edgartown and Oak Bluffs, and obituaries.

25. For Henry Fletcher Worth and the necessary reminder of the constancy and intensity of Vineyard ties to the distant Pacific all through the nineteenth century, see Emma Mayhew Whiting and Henry Beetle Hough, *Whaling Wives* (Boston: Houghton Mifflin, 1953). Charles Worth's obituary is in the *Cottage City Star,* 23 January 1884.

26. *Gazette,* 16 March and 13 July 1866, 8 May 1868, and 23 December 1870; *Standard,* 18 August 1868; Vincent (1870) p. 146.

27. *Gazette,* 8 May 1868.

28. *Gazette,* 28 April and 28 May 1871; *Cottage City Star,* 6 and 27 May 1880.

29. *Gazette,* 14 July 1867.

CHAPTER FOUR

1. Vincent (1870), pp. 127–129, 180.

2. *Zion's Herald,* 23 August 1865; *Standard,* 28 August 1868.

3. Vincent (1870), pp. 133–134.

4. *Standard,* 20 and 21 August, 1863; *Journal,* 18 August 1863; *Camp Meeting Herald,* 21 August 1866. With the focus on cottages, almost no mention was made through the years of the old society or "mammoth" tents for church groups. These lived on at Wesleyan Grove, serving as rainy-day chapels and as hostelries, with rows of men and women sleeping in straw on separate sides of the canvas structures. A *New York Times* correspondent in 1867 (August 20) bedded at one of the four New Bedford tents, buying a ticket for the purpose. There were pegs around the sides for clothing, rows of trunks and settees as furniture, and a small dressing and washroom in the rear. According to the protagonist of Antoinette Brown Blackwell's fictionalized account of a trip to Wesleyan Grove (*The Island Neighbors,* New York, 1871, pp. 91–92), it was tough sleeping in these quarters because of the heat and others' restlessness. New society tents were built in 1871, but they declined, as a whole, from 40 in 1869 to 15 in 1878, the last being removed in 1883. At other northeastern campgrounds society tents developed into wooden buildings with meetings and dining on the first floor and lodging on the second. A New York group proposed such a "chapel" in the rustic style for Wesleyan Grove, but nothing came of the idea. (*Journal,* 26 August 1860).

5. *Gazette,* 24 June 1864, 24 and 31 August 1866.

6. *Camp Meeting Herald,* 22 August 1866.

7. *The New York Times,* 31 August 1866, 20 August 1867; *Gazette,* 31 August 1866.

8. *Standard,* 18 August 1866 and 27 August 1871.

9. Gilbert Haven, *Father Taylor, The Sailor Preacher* (New York: Phillips and Hunt, 1881), pp. 228, 298.

10. *Fraternal Camp-Meeting Sermons* (New York: Nelson and Phillips, 1875), pp. 228, 459–496.

11. N. Robinson, "The Islands of the Bay State," *Frank Leslie's Popular Monthly,* 12(1), July 1881, p. 103; "Camp-Meeting at Martha's Vineyard," *Harper's Weekly,* XII(611), 12 September 1868, p. 580.

12. *Gazette,* 20 August 1869; *The New York Times,* 20 August 1866.

13. *Standard,* 13 August 1864.

14. Robinson, pp. 102–103.

15. Almer M. Pennewell, *The Methodist Movement in Northern Illinois* (Sycamore: *Sycamore Tribune,* n.d.) pp. 275, 287–288; *The New York Times,* 31 August 1866; *Standard,* 10 August 1865, 16 August 1867; Vincent (1870), p. 182.

16. James Jackson Jarves, "A New Phase of Druidism," *Galaxy,* 10(6), December 1870, pp. 777–783.

17. Quoted by Charles A. Johnson, *The Frontier Camp Meeting* (Dallas: Southern Methodist University Press, 1955), p. 94.

CHAPTER FIVE

1. Vincent (1870), p. 46.

2. The history of Oak Bluffs is extensively described in Hough, upon which this account relies.

3. *Gazette,* 16 November 1866.

4. A recent Massachusetts law, Chapter 59, 1867, protected camp meetings from commercial intrusion. See *Acts and Resolves Passed by the General Court of Massachusetts in the Year 1867* (Boston: Wright and Potter, 1867); *Standard,* 10 and 19 August 1867.

5. *Journal,* 15 August 1870; Vincent (1870), p. 247; *Standard,* 24 August 1871.

6. *Journal,* 21 August 1872. For Llewellyn Park see Jane B. Davies, "Llewellyn Park in West Orange, New Jersey," *Antiques Magazine* 107, January 1975, pp. 142–155; and Richard Guy Wilson, "Idealism and the Origin of the First American Suburb: Llewellyn Park, New Jersey," *The American Art Journal,* XI, October 1979, pp. 79–90.

7. For Riverside, see Walter L. Creese, *The Crowning of the American Landscape: Eight Great Spaces and Their Buildings* (Princeton: Princeton University Press, 1985) pp. 219–240. The Le Vesinet connection is in Theodore Turak, "Riverside Roots in France," *Inland Architect,* 25(9), November–December 1981. An argument for English sources for the American romantic suburb can be found in John Archer, "Country and City in the American Romantic Suburb," *JSAH,* 42(2), May 1983.

8. For Eastus P. Carpenter, see Hough, pp. 62–63; John B. Hodges, "History of Foxboro XXXIV," *Foxboro Reporter,* 17 November 1925; and Mrs. Clifford W. Lane, *This Was Foxborough* (Foxborough: Croft Press, 1966).

9. A history of Oaks Bluffs's plans and the Copeland biography are in Ellen Weiss, "Robert Morris Copeland's Plans for Oak Bluffs," *JSAH,* 34(1), March 1975.

10. Chris Stoddard, *A Centennial History of Oak Bluffs* (Oak Bluffs Historical Commission, 1981), p. 47.

11. *Standard,* 2 August 1867, 17 and 24 August 1868; John Stevens's cottage for E. P. Carpenter was shown on page 113 of John B. Bachelder, *Popular Resorts and How to Reach*

Them, (Boston: J. B. Bachelder, 1874). Copies of the 1878 lithograph of the first Oak Bluffs building are at DCHS, the offices of the *Gazette,* and the Boston Athenaeum.

12. The *Gazette* claimed 30 new cottages in the area on July 2, 1869, and 60 in Oak Bluffs alone by·August 5.

13. The basis of attribution of Oak Bluffs's buildings to Pratt is discussed in Ellen Weiss, "Introducing S. F. Pratt," *Nineteenth Century,* 4(3), Autumn 1978. Since this article, newspaper attributions to Pratt of Union Chapel and the Sea View Hotel have been found.

14. For Robert Morris Copeland and S. F. Pratt on Shelter Island (Suffolk County), see the New York State cultural resources inventory forms of 1979 by Ellen Cole, Ellen Williams, and the Society for Preservation of Long Island Antiquities.

15. For a thorough exploration of the European background of the Stick Style as a rustic mode, see Sarah Bradford Landau, "Richard Morris Hunt, the Continental Picturesque, and the 'Stick Style,'" *JSAH,* 42(3), October 1983, pp. 272–289. For Haiti, see Anghelen Arrington Phillips, Jr., *Gingerbread Houses: Haiti's Endangered Species,* Port-au-Prince, 1977.

16. *Standard,* 23 August 1871. Carpenter's own cottage was an earlier one, designed by John Stevens.

17. James Mudge, *History of the New England Conference of the Methodist Episcopal Church* (Boston: The Conference, 1910), p. 261–263.

18. For Pratt's Newport house, "Birds Nest Cottage," see Antoinette F. Downing and Vincent J. Scully, Jr., *The Architectural Heritage of Newport, R.I.* (Cambridge: Harvard University Press, 1952), plate 181; also see Weiss (1978).

19. The best published survey of Pratt's Newport sources is George C. Mason, *Newport and its Cottages,* (Boston: J. R. Osgood, 1875). For George C. Mason's Mrs. Loring Andrews's house, dated 1872 by Scully, see Downing and Scully, p. 135. The exact relations between Pratt and Mason await analysis.

20. *Standard,* 18 August 1871.

21. Hough, p. 78. A lengthy description of the Sea View on opening is in the *Journal,* 26 July 1872; another description commemorating its cremation is in the *Gazette,* 29 September 1892. For the possible relation between John Stevens and the earliest Sea View scheme, see note 15. The last owner of the hotel published a design for a Shingle Style reconstruction about 1895, but nothing came of the idea. See Frederic J. Hart, *Lovers Rock, A Summer Idyll of Cottage City, Mass.* (New York: South Publishing Co., [c. 1895]).

22. *Boston Globe,* 27 August 1875.

23. *Gazette,* 5 May 1961.

24. Hough, pp. 194–195; *Cottage City Star,* 23 March 1882; *Gazette,* 20 April 1888 and 5 May 1950.

CHAPTER SIX

1. *Standard,* 31 August 1868; *Journal,* 31 July 1869, 5 September 1874. According to the *Journal* for 23 August 1878, J. A. S. Monk was in Oak Bluffs making drawings for the bird's-eye view published by Sunderland of Providence in 1880. The *Gazette* of 15 August 1879 reported that a Mr. White revised the bird's eye to show the tabernacle, the rink, and the wharf built between the 1878 drawing and the 1880 publication. The 1914 figure is from a Sanborn Insurance Company map made in December.

2. *Journal,* 5 September 1874; *A Guide to Martha's Vineyard and Nantucket* (Boston: Rockwell and Churchill, 1876).

3. Hough, opposite p. 148.

4. *Journal,* 19 August 1878.

5. For Isaac Rich, see James Mudge, *History of The New England Conference of the Methodist Episcopal Church* (Boston: The Conference, 1910); and *Standard,* 2 September 1871.

6. *Argument of Charles R. Train Before the Committee on Towns* (Boston: 1879), n.p.

7. *Standard,* 11 August 1871.

8. *A Guide to Martha's Vineyard and Nantucket,* pp. 23, 41; Arthur E. Sproul, *Tourists Guide to Nantasket Beach, Downer Landing, Martha's Vineyard, Nantucket . . .* (Boston: Everett and Zerrahn, 1882), pp. 27–28; N. Robinson, "The Islands of the Bay State," *Frank Leslie's Popular Monthly* 12(1), July 1881, p. 99.

9. Quoted in Hough, p. 129; Charles Dudley Warner, "Their Pilgrimage," *Harper's New Monthly Magazine* 63(434), July 1886, pp. 172–174.

10. *Journal,* 8 August 1872, 2 August 1875; Rev. Frederic Denison, *Illustrated New Bedford, Martha's Vineyard, and Nantucket,* (Providence, 1880), n.p. The date of Thomas Hughes's visit to the Vineyard is furnished by the *Journal* for July 16, 1875. For Rugby Tennessee, see Lee R. Minton, "Rugby, Tennessee: Thomas Hughes's 'New Jerusalem,'" *Nineteenth Century,* 7(4), Winter–Spring 1982, pp. 43–46.

11. Hough, pp. 131–132; Nathaniel Southgate Shaler, "Martha's Vineyard,"*The Atlantic Monthly,* December 1874, pp. 732–740.

12. Andrew Dickson White, *Autobiography of Andrew Dickson White* (New York: The Century Co., 1905) II, pp. 389–390.

13. Hough, pp. 132–138; *Standard,* 16 August 1869; *Journal,* 10, 15, 16 August 1872; 20 August 1873; 18 August 1874.

14. *Journal,* 20 and 22 August 1877.

15. *Journal,* 20 August 1871.

16. Hough, pp. 139–144; *Journal,* 18 August 1874 and 14 August 1869.

17. *Journal,* 20 August 1870, 29 July and 13 August 1872, and 15 August 1873.

18. *Journal,* 8 August 1872.

19. D. H. Strother, "A Summer in New England II," *Harper's New Monthly Magazine,* XXI, September 1860, pp. 442–461.

20. *Standard,* 23 August 1869; Vincent (1870), pp. 228–229.

21. For the National Holiness Camp Meeting Association see George Hughes, *Days of Power in the Forest Temple* (Boston: John Bent, 1873). We have already seen that the Perfectionist advocate and camp-meeting manual author, Rev. B. W. Gorham, was not a favorite at Wesleyan Grove by 1865. The famed New York Perfectionist speakers Dr. and Mrs. Palmer came to the Vineyard in 1865 and were described with considerable reserve. Both husband and wife pronounced their "peculiar statements" for the meeting, but the audience was pleased to hear them anyway and to "study the secret of their power over the people." The Palmers led and overdid the Love Feast on that visit (*Standard,* 22 August 1865).

22. *Gazette,* 7 August 1936.

23. This account combines those from two sources: Ira W. LeBaron, *The Campground at Martha's Vineyard* (Nashville: Parthenon Press, [1958]), pp. 34–35; and George Prentice, *The Life of Gilbert Haven* (New York: Phillips and Hunt, 1883), pp. 386–387. For more on Grant's visit see Hough, pp. 113–116, and the *Journal* for 28, 29, 31 August and 1 September 1874.

24. *Gazette,* 23 August 1935.

25. For the skating rink, see *The Cottage City, or The Season at Martha's Vineyard* (Lawrence, Mass.: S. Merrill and Crocker, 1879). p. 27, and Chris Stoddard, *A Centennial History of Cottage City* (Oak Bluffs: Oak Bluffs Historical Commission, 1981), p. 90.

26. For the 1874 plan, *Seaside Gazette,* 4 August 1874. On June 9, 1877 plans were revealed for a building to seat 600, which was to stand within the preaching area, facing out to Broadway (*Newport Journal,* 9 June 1877). Design for the present chapel was accepted that August. It was placed where the County Street, Fourth Street, and Chilmark society tents had stood, facing in toward the preaching area. Construction was underway in December, clapboarding in February 1878, and painting and frescoing of the interior by Clark and Berthold in May.

27. Edward Lawrence Hyde, born in Mystic, Connecticut, in 1835 and converted in 1859, studied architecture and then art in New York City with Johannes A. Oertel of the National Academy. He was called to the ministry in 1868 and studied at Boston University (Miller, III, p. 266).

28. The history and chronology of the tabernacle is taken from Ellen Weiss, "The Iron Tabernacle at Wesleyan Grove," *The Dukes County Intelligencer,* 21(1), August 1979. Sources were the *Record of the Martha's Vineyard Camp Meeting Association,* 1860, Book No. 2, pp. 161–165, which the directors of the association kindly permitted me to consult, and *The Vineyard Gazette, The Cottage City Star, The Springfield Republican, The New Bedford Evening Standard,* and *The New Bedford Daily Mercury.* Material from *The Providence Daily Journal* was found after this article was published.

There has been some hope among the tabernacle's admirers that it is a rare early steel structure. Steel manufacture in 1879 at architectural scale was limited to specially made and therefore prohibitively expensive pieces, not stock building components. The principal supplier for this building, the Phoenix Iron Co., near Philadelphia, did not list steel sections in its catalog for another seven years.

29. A campground tradition holds that the wooden cupola made in Germany for use at the Philadelphia Exposition of 1876, was not used there, and was added to this project. This has not been substantiated. There is indication, however, from an early description, probably taken from the contract or drawings, that the cupola was not part of the original plan. The description calls the area under the top roof, the point 75 feet up, the "dome," and refers to a flagpole that will go up from the dome to attain 100 feet (*Standard,* 19 May 1879).

30. The only Hoyt biographical information is in David W. Hoyt, *A Genealogical History of the Hoyt, Haight, and Hight Families* (Providence: 1871), p. 467. According to the 1931 *Alumni Record of Wesleyan University,* he died in July 1911, place unrecorded.

31. The Rev. Frederic Denison, *Illustrated New Bedford, Martha's Vineyard, and Nantucket* (Providence: 1880); *Journal,* 8 August 1879.

32. Walter Prichard Eaton, *Martha's Vineyard, A Pleasant Island in a Summer Sea* (New York: 1930), n.p.

33. Adelaide M. Cromwell, "The History of Oak Bluffs as a Popular Resort for Blacks," *The Dukes County Intelligencer,* 21(1), August 1984, pp. 3–25.

CHAPTER SEVEN

1. *Camp Meeting Herald,* 22 August 1866.

2. Gilbert Haven, "John Ruskin," *The Methodist Quarterly Review* (Fourth Series), 12(34), October 1860.

3. Excerpted in the *Gazette,* 23 June 1967, from the Martha's Vineyard Summer Institute *Herald,* July 1882.

4. Robert Rosenblum, "The Primal American Scene," in *The Natural Paradise: Painting in America, 1800–1950* (New York: The Museum of Modern Art, 1976), pp. 14–37.

5. Peter Cartwright, *Autobiography of Peter Cartwright, The Backwoods Preacher,* (Cincinnati: Cranston and Curtis n.d.), p. 476.

6. *The National Magazine,* VII, December 1855, pp. 497–512.

7. Vineland, founded in 1861 by Charles K. Landis in western New Jersey, attracted Methodists with its temperance regulations. The colony was organized into small market-garden farms of 20 to 60 acres in a deliberate attempt to provide villagelike human density for the residents, and thus social, mental, and religious culture far from the evils of real cities. The plan of the development was a square of a mile on each side. Aesthetic features included rows of shade trees bordering the railroad track, a policy to discourage fencing, fruit and ornamental trees on public ways, and 20-foot setbacks for houses, all consciously set forth to achieve order, beauty, and morality. Better-known products of the colony are Dr. Thomas B. Welch's grape juice preservation methods and John C. Mason's glass jars. See Philip Snyder, "Vineland as a Lesson in Colonization," *U.S. Department of Agriculture Report 1869–1870,* pp. 410–415, and Charles K. Landis, *The Founder's Own Story* (Vineland: Vineland Historical and Antiquarian Society, 1903), and David P. Handlin, *The American Home* (Boston: Little, Brown, 1979).

8. Robert Winston Taylor, *The Search for a New Urban Order: The Vision of a Suburban Civilization,* Ph.D. thesis, 1971 (St. Louis: University Microfilms, 1972).

9. Walter L. Creese, "Imagination in the Suburb," in U. C. Knoepflmacher and G. B. Tennyson (Eds.) *Nature and the Victorian Imagination* (Berkeley: University of California Press, 1977), pp. 49–67.

BIBLIOGRAPHIC ESSAY

THE HISTORY of Wesleyan Grove and Oak Bluffs has been told several times, always with full appreciation for the extraordinary human events although without the environmental analysis which is the focus of the present book. The Reverend Hebron Vincent, secretary of the camp meeting for 19 years, wrote two volumes appearing in 1858 and 1870. These volumes constitute the only extensive history of any Methodist camp meeting. Henry Beetle Hough, editor of *The Vineyard Gazette* from 1920 until his death in 1985, retold Vincent's story in 1936 with embellishments from other sources, including his own memories, in *Martha's Vineyard, Summer Resort.* Hough showed how the solemn religious tone of the camp meeting, and even its unruly festive outcroppings, were eclipsed at Oak Bluffs by the more ebullient style and extroverted mood of the resort. He presents the political and financial shenanigans, the ambitious and energetic inhabitants, and the rituals of pleasure and communal joy. Hough knew Oak Bluffs as a boy at the beginning of our century and, as a young editor, interviewed campground residents whose memories reached back to the 1850s. His book must be taken as precise history and a superb imaginative recreation of a place and an age. More recently, Christine Stoddard has written another history for the occasion of Oak Bluffs's centennial, an account that is fresh, wise, well illustrated, and priced for easy purchase.

The earlier twentieth century has also produced a number of brief published observations of Wesleyan Grove and Oak Bluffs which singled out different aspects of the extraordinary visual quality of the communities. Talbot Hamlin noted the architecture of Oak Bluffs in 1938 in a bold attempt to reconcile his anti-Victorian convictions with the impressive formal consistency of the town. Victorian architecture, to Hamlin, mirrored the over-rapid exploitation and cutthroat competition of the era, its "forced ostentation [and] over-lavish and ill considered detail," but he also found that Oak Bluffs had a "queer charm of its own" and that there was an "exuberant delight in wood for its own sake, worked with all the available tools and made into a foundation for an extraordinary dreamlike fantasy." At about the same time, Aaron Siskind, then a vacationing high school teacher, photographed the campground and exhibited the results in New York. His aim was to evoke the atmosphere of the earlier era and to reveal the beauty in Victorian survivals, for, to quote him, "they are beautiful in their own special way." In the early and middle 1950s, Carroll L. V. Meeks, professor of architectural history at Yale and a Martha's Vineyard summer resident, introduced members of the Society of Architectural Historians to the community. *Gazette* editorials relayed their professional opinion that the town ought to be a historic district on the model of Charleston. Photographs of the jigsaw work appeared in *Architectural Forum* in May 1960. Historic photographs were published by another island-based historian of American life, David G. McCullough, in *American Heritage* in October 1967.

The author of the present study started work in 1973 with a lecture for the joint session of the American and British societies of architectural historians meeting in Cambridge, England. This material was repeated for a variety of New England audiences and was published in *Architecture Plus* in November 1973.

In 1974 the community appeared briefly in *The Place of Houses* by Charles Moore, Donlyn Lyndon, and Gerald Allen. The authors admired the "paleo-Disneyland" as an evocation of fairytale forests and the cottages for being "winsomely funny" rather than merely "silly." Other recent appearances in an architectural context include William H. Pierson's *American Buildings and Their Architects,* in which Wesleyan Grove was characterized as "one of the most remarkable concentrations of folk architecture anywhere in the country." Dolores Hayden considered the cottages briefly in *The Grand Domestic Revolution.* Wesleyan Grove and Oak Bluffs were included within the survey of English and American suburbs in a 1981 issue of the British periodical, *Architectural Design.* Most recently, William Nathaniel Banks has written an excellent introduction to the history of the camp meeting in *The Magazine Antiques.* Wesleyan Grove was listed in the National Register of Historic Places in 1978.

BIBLIOGRAPHY

Acts and Resolves Passed by the General Court of Massachusetts in the Year 1867. Boston: Wright and Potter, 1867.

Adams, John. *The Life of "Reformation" John Adams*. Boston: George C. Rand, 1853.

Ahlstrom, Sidney E. *A Religious History of the American People*. New Haven: Yale University Press, 1972.

Allen, Stephen. *The Life of Rev. John Allen, Better Known as "Camp Meeting John."* Boston: B. B. Russell, 1888.

Alumni Record of Wesleyan University. Third Edition. Hartford: Case Lockwood and Brainard, 1883.

American Guide Series. *New Jersey, A Guide to the Present and the Past*. New York: Viking Press, 1939.

Archer, John. "Country and City in the American Romantic Suburb." *Journal of the Society of Architectural Historians* 42, May 1983, 139–156.

Argument of Charles R. Train Before the Committee on Towns. Boston: 1879, n.p.

Bachelder, John B. *Popular Resorts and How to Reach Them*. Boston: J. B. Bachelder, 1875.

Baker, Anne W. "The Vincent House: Architecture and Restoration." *The Dukes County Intelligencer* 20, August 1978, 7–27.

Baker, George Claude. *An Introduction to the History of Early New England Methodism*. Durham: Duke University Press, 1941.

Bangs, Nathan. *A History of the Methodist Episcopal Church*. New York: T. Mason and G. Lane, 1839.

Banks, Charles E. *History of Martha's Vineyard*. Boston: George H. Dean, 1911.

Banks, William Nathaniel. "The Wesleyan Grove Campground on Martha's Vineyard." *The Magazine Antiques*, July 1983, 104–116.

Blackwell, Antoinette Brown. *The Island Neighbors*. New York: Harper and Bros. 1871.

The Book of Discipline of the United Methodist Church. Nashville: The United Methodist Publishing House, 1976.

Bruce, Dickson D., Jr. *And They All Sang Hallelujah*. Knoxville: The University of Tennessee Press, 1974.

Cameron, Richard M. *Methodism and Society in Historical Perspective*. New York: Abingdon Press, 1961.

"Camp-Meeting at Martha's Vineyard" *Harper's Weekly* XII, September 12, 1868, 580.

Candee, Richard M. "A Documentary History of Plymouth Colony Architecture, 1620–1700: Construction Methods." *Old-Time New England* LX, October–December 1969, 36–53.

Cartwright, Peter. *Autobiography of Peter Cartwright, The Backwoods Preacher.* Cincinnati: Cranston and Curtis, [n.d.].

Carwardine, Richard. *Transatlantic Revivalism: Popular Evangelism in Britain and America, 1790–1865.* London: Greenwood Press, 1978.

Chamberlain, J. N. *Cottage City Illustrated.* Woonsocket, RI: 1888.

Clark, David. *The Way Rev. Moses L. Scudder Secured a Cottage at Martha's Vineyard.* Hartford: 1870.

Clements, William M. "The Physical Layout of the Methodist Camp Meeting." *Pioneer America* 5, January 1973, 9–15.

Coffin, Sirson P. *Annual Report of the Agent of the Martha's Vineyard Camp-Meeting.* New Bedford: 1860.

Connally, Ernest Allen. "The Cape Cod House: An Introductory Study." *Journal of the Society of Architectural Historians* XIX, May 1960, 47–56.

The Cottage City, or The Season at Martha's Vineyard. Lawrence, Mass.: S. Merrill and Crocker, 1879.

Creese, Walter L. *The Crowning of the American Landscape: Eight Great Spaces and Their Buildings.* Princeton: Princeton University Press, 1985.

Creese, Walter L. "Imagination in the Suburb," U. C. Knoepflmacher and G. B. Tennyson, (Eds.) *Nature and the Victorian Imagination.* Berkeley: University of California Press, 1977, 49–67.

Cromwell, Adelaide M. "The History of Oak Bluffs as a Popular Resort for Blacks," *The Dukes County Intelligencer* 26(1), August 1984, 3–25.

Dagnall, Sally W. *Martha's Vineyard Camp Meeting Association, 1835–1985.* Oak Bluffs: The Association, 1984.

Davies, Jane B. "Llewellyn Park in West Orange, New Jersey." *The Magazine Antiques* 107, January 1975, 142–155.

Denison, Frederic. *Illustrated New Bedford, Martha's Vineyard, and Nantucket.* Providence: 1880.

[Devens, Samuel Adams.] *Sketches of Martha's Vineyard, and Other Reminisences of Travel at Home, Etc.* Boston: James Munroe and Co., 1838.

Downing, Antoinette F. *Early Homes of Rhode Island.* Richmond: Garrett and Massie, 1937.

Downing, Antoinette F., and Vincent J. Scully, Jr. *The Architectural Heritage of Newport, R.I.* Cambridge: Harvard University Press, 1952.

Eaton, Walter Prichard. *Martha's Vineyard, A Pleasant Island in a Summer Sea.* New York, New Haven and Hartford Railroad and New England Steamship Co., 1923.

Fraternal Camp-Meeting Sermons. New York: Nelson and Phillips, 1875.

Ford, Robert M. *Mississippi Houses: Yesterday Toward Tomorrow.* State College: privately printed, 1982.

Gookin, Warner F. "The Jordan Road and Oak Bluffs Harbor." *Oak Bluffs Leaflet No. 1,* n.d.

Gorham, B. W. *Camp Meeting Manual, A Practical Book for the Camp Ground.* Boston: H. V. Degen, 1854.

A Guide to Martha's Vineyard and Nantucket. Boston: Rockwell and Churchill, 1876.

Hamlin, Talbot F. "Americana, Spirit of Early Building Transcends Periods." *Pencil Points* 19, October 1938, 655–662.

Hamlin, Talbot F. *Benjamin Henry Latrobe.* New York: Oxford University Press, 1955.

Handlin, David P. *The American Home.* Boston: Little, Brown, 1979.

Hart, Frederic J. *Lovers Rock, A Summer Idyl of Cottage City, Mass.* New York: South Publishing Co., [1895].

Haven, Gilbert. *Father Taylor, The Sailor Preacher.* New York: Phillips and C. Hunt, 1871.

Haven, Gilbert. "John Ruskin." *The Methodist Quarterly Review,* Fourth Series XII, October 1860, 533–554.

Hayden, Dolores. *The Grand Domestic Revolution.* Cambridge: MIT Press, 1981.

Hazen, Joseph C. "Jigsaw City." *Architectural Forum* 112, May 1960, 134–139.

Hine, C. G. *The Story of Martha's Vineyard.* New York: Hine Bros., 1908.

Hodges, John B. "History of Foxboro XXXIV." *Foxboro Reporter,* 17 November 1925.

Hough, Henry Beetle. *Martha's Vineyard, Summer Resort.* Rutland, Vt.: The Tuttle Publishing Co., 1936.

Hoyt, David W. *A Genealogical History of the Hoyt, Haight, and Hight Families.* Providence: 1871.

Hughes, George. *Days of Power in the Forest Temple.* Boston: John Bent, 1873.

Iarocci, Joseph J. "The Cottage City of America: Secularization and Urbanization in the Evolution of a Resort Community." Unpublished senior thesis, Department of History, Brown University, 1981.

Jackson, John Brinckerhoff. *The Necessity for Ruins.* Amherst: The University of Massachusetts Press, 1980.

Jarves, James Jackson. "A New Phase of Druidism." *The Galaxy* 11, December 1870, 777–783.

Johnson, Charles A. *The Frontier Camp Meeting.* Dallas: Southern Methodist University Press, 1955.

Jones, Charles Edwin. *Perfectionist Persuasion: The Holiness Movement and American Methodism, 1867–1936.* Metuchen, NJ: The Scarecrow Press, 1974.

Lacossitt, Henry. "Yankee Camp Meeting." *Saturday Evening Post,* August 20, 1955.

Landau, Sarah Bradford. "Richard Morris Hunt, the Continental Picturesque, and the 'Stick Style.'" *Journal of the Society of Architectural Historians* 42, October 1983, 272–289.

Landis, Charles K. *The Founder's Own Story.* Vineland: Vineland Historical and Antiquarian Society, 1903.

Lane, Mrs. Clifford W. *This Was Foxborough.* Foxborough: Croft Press, 1966.

LeBaron, Ira W. *The Campground at Martha's Vineyard.* Nashville: Parthenon Press, [1958].

Lee, Jesse. *A Short History of the Methodists in the United States of America.* Baltimore: Magill and Clime, 1810.

Lobeck, A. K. *A Brief History of Martha's Vineyard Camp-Meeting Association.* Oak Bluffs: 1956.

Ludlum, Stuart D. *Exploring Nantucket, Martha's Vineyard 100 Years Ago.* Utica: Brodock and Ludlum Publications, 1973.

McCullough, David G. "Oak Bluffs." *American Heritage* 12 October 1967, 65–70.

McFerrin, John B. *History of Methodism in Tennessee.* Nashville: Southern Methodist Publishing House, 1871.

McLoughlin, William G. (Ed.) *The American Evangelicals, 1800–1900.* New York: Harper and Row, 1968.

McLoughlin, William G. *Modern Revivalism.* New York: Ronald Press, 1959.

Martha's Vineyard and its Attractions. New York: Geo. W. Richardson, [n.d.].

Martha's Vineyard: Its History and Advantages as a Health and Summer Resort. Providence: Freeman and Son, 1889.

Mason, George Champlin. *Newport and its Cottages.* Boston: J. R. Osgood, 1875.

Mayer, F. E. *The Religious Bodies of America.* St. Louis: Concordia Publishing House, 1958.

Meeks, Carroll L. V. "Romanesque Before Richardson in the United States." *Art Bulletin* 35, March 1953, 17–33.

"Methodist Church Architecture." *The National Magazine* VII, December 1855, 497–512.

Miller, Rev. Rennetts C. *Souvenir History of the New England Southern Conference.* Nantasket, Mass.: 1897.

Milner, Vincent L. *Religious Denominations of the World.* Philadelphia: Bradley and Company, 1872.

Minton, Lee R. "Rugby, Tennessee: Thomas Hughes's 'New Jerusalem.'" *Nineteenth Century* 7, Winter–Spring 1982, 43–46.

Moore, Charles, Gerald Allen, and Donlyn Lyndon. *The Place of Houses.* New York: Holt, Rinehart and Winston, 1974.

Mudge, James. *History of the New England Conference of the Methodist Episcopal Church.* Boston: The Conference, 1910.

Nichols, Thomas L. *Forty Years of American Life.* Vol. I. London: John Maxwell and Co., 1864.

Nordhoff, Charles. "Cape Cod, Nantucket, and the Vineyard." *Harper's New Monthly Magazine* LI, June 1875, 52–66.

Norton, Henry Franklin. *Martha's Vineyard.* 1923.

Nottingham, Elizabeth K. *Methodism and the Frontier: Indiana Proving Ground.* New York: Columbia University Press, 1941.

"Oak Bluffs." *Architectural Design* 51, October–November 1981, 66–67.

"Oak Bluffs Revived," *Massachusetts Historical Commission Newsletter* 7, March–April 1981, 2–3.

Parker, Charles A. "The Camp Meeting on the Frontier and the Methodist Religious Resort in the East—Before 1900." *Methodist History* 18,3, April 1980.

Parker, Charles A. "Ocean Grove N.J.: Queen of the Victorian Methodist Camp Meeting Resorts." *Nineteenth Century* 9, 1984, 19–25.

Pease, Jeremiah. "The Island's First Methodists." *The Dukes County Intelligencer* 22, November 1980, 58–70.

[Pease, Richard L., and Sumner Myrick]. *A Guide to Martha's Vineyard and Nantucket with a Directory of the Campground.* Boston: Rockwell & Churchill, 1876.

Pennewell, Almer M. *The Methodist Movement in Northern Illinois.* Sycamore, Il.: *Sycamore Tribune* [n.d.].

Phillips, Anghelen Arrington, Jr. *Gingerbread Houses: Haiti's Endangered Species.* Port-au-Prince: 1977.

Pierson, William H., Jr. *American Buildings and Their Architects: Technology and the Picturesque, the Corporate and Early Gothic Styles.* Garden City: Doubleday, 1978.

[Porter, James]. *An Essay on Camp-meetings.* New York: Lane and Scott, 1849.

Port-folio of Views of Cottage City, Martha's Vineyard, Mass. Boston: Baldwin Coolidge and A. N. Houghton, 1886.

Prentice, George. *The Life of Gilbert Haven.* New York: Phillips and Hunt, 1883.

[Railton, Arthur R., editorial note] *The Dukes County Intelligencer,* May 1985, 184.

Railton, Arthur R. "Jeremiah's Travail." *The Dukes County Intelligencer,* November 1980, 43–57.

Reps, John W. *The Making of Urban America.* Princeton: Princeton University Press, 1965.

Robinson, N. "The Islands of the Bay State." *Frank Leslie's Popular Monthly* 12, July 1881.

Rosenblum, Robert. "The Primal American Scene," in *The Natural Paradise: Painting in America, 1800–1950.* New York: The Museum of Modern Art, 1976, 14–37.

Roth, Leland M. *A Concise History of American Architecture.* New York: Harper and Row, 1980.

Schlereth, Thomas J. "Chautauqua: A Middle Landscape of the Middle Class." *Henry Ford Museum and Greenfield Village Herald* 13,2, 1984 22–31.

Senter, Oramel S. "Oak Bluffs." *Potter's American Monthly* IX, August 1877; reprinted in *The Dukes County Intelligencer* 21, May 1980.

Shaler, Nathaniel Southgate. "Martha's Vineyard." *The Atlantic Monthly,* December 1874, 732–740.

Shoppell, Robert W. *How to Build, Furnish, and Decorate.* New York: The Cooperative Building Plan Association, 1883.

Simpson, Bishop Matthew. *Cyclopaedia of Methodism.* Philadelphia: Everts and Stewart, 1878.

Smeltzer, Wallace Guy. *Methodism in the Headwaters of the Ohio.* Nashville: The Parthenon Press, 1951.

Smith, Elizabeth Simpson. "Camp Meeting Time." *Historic Preservation* 30, April–June 1978, 20–23.

Smith, Timothy L. *Revivalism and Social Reform.* Nashville: Abingdon Press, 1957.

Snyder, Philip. "Vineland as a Lesson in Colonization." *U.S. Department of Agriculture Report 1869–1870,* 410–415.

Sproul, Arthur E. *Tourists Guide to Nantasket Beach, Downer Landing, Martha's Vineyard, Nantucket* . . . . Boston: Everett and Zerrahn, 1882.

Stansfield, Charles. "Pitman Grove: A Camp Meeting as Urban Nucleus." *Pioneer America* 7, January 1975, 36–44.

Stilgoe, John R. *Common Landscape of America, 1580–1845.* New Haven: Yale University Press, 1982.

Stoddard, Chris. *A Centennial History of Cottage City.* Oak Bluffs: Oak Bluffs Historical Commission, 1981.

Strother, D. H. "A Summer in New England II." *Harper's New Monthly Magazine* XXI, September 1860, 442–461.

Supreme Judicial Court. *Attorney General, Ex.rel., vs. George C. Abbott.* Boston, 1890.

Sweet, William Warren. *Methodism in American History.* New York: The Methodist Book Concern, 1933.

Sweet, William Warren. *Story of Religions in America.* New York: Harper and Bros., 1930.

Sweetser, Charles H. *Book of Summer Resorts.* New York: J. R. Osgood, 1868.

Taylor, Robert Winston. *The Search for a New Urban Order: The Vision of a Suburban Civilization.* Ph.D. Thesis, 1971, St. Louis University; University Microfilms, 1972.

Tebbetts, Diane. "Traditional Houses of Independence County, Arkansas." *Pioneer America* 10, 1 June 1978, 36–55.

de Tocqueville, Alexis. *Democracy in America.* New York: Harper and Row, 1966.

The Tourists' Guide to Southern Massachusetts, Camp Meeting Edition. New Bedford: Taber Brothers, 1868.

Trachtenberg, Alan. "'We Study the Word and Works of God'; Chautauqua and the Sacralization of Culture in America." *Henry Ford Museum and Greenfield Village Herald* 13, 2, 1984, 3–11.

[Train, Charles R.] *Argument of Charles R. Train Before the Committee on Towns.* Boston: 1879.

Trollope, Frances. *Domestic Manners of the Americans.* London: 1832.

Turak, Theodore. "Riverside Roots in France." *Inland Architect* 25, November–December 1981.

Vincent, Hebron. *A History of the Camp Meeting and Grounds at Wesleyan Grove, Martha's Vineyard, for the Eleven Years Ending With the Meeting of 1869.* Boston: Lee, 1870.

Vincent, Hebron. *History of the Wesleyan Grove Camp Meeting from the First Meeting Held There in 1835 to That of 1858*. Boston: George C. Rand and Avery, 1858.

Vincent, Hebron. "Men of the Past—Brief Memories."*Zion's Herald*, April 22, 1885.

[Vincent, Hebron.] *The Vineyard as it Was, Is, and Is To Be, By an Observer*. New Bedford: 1872.

Warner, Charles Dudley. "Their Pilgrimage." *Harper's New Monthly Magazine* LXIII, July 1886, 172–174.

Weisberger, Bernard. *They Gathered at the River*. Boston: Little, Brown, 1958.

Weise, Arthur James. *History of Round Lake*. Troy, N.Y.: Douglas Taylor, 1887.

Weiss, Ellen. "Introducing S. F. Pratt." *Nineteenth Century* 4, Autumn 1978, 89–93.

Weiss, Ellen. "The Iron Tabernacle at Wesleyan Grove." *The Dukes County Intelligencer* 21, 1979, 3–14.

Weiss, Ellen. "Robert Morris Copeland's Plans for Oak Bluffs."*Journal of the Society of Architectural Historians* 34, March 1975, 60–66.

Weiss, Ellen. "Samuel Freeman Pratt: An Architect of Oak Bluffs." *The Dukes County Intelligencer* 21, May 1980, 123–132.

Weiss, Ellen. "The Wesleyan Grove Campground." *Architecture Plus* I, November 1973, 44–49.

When Oak Bluffs was Cottage City. Providence: Henry P. Porter Publisher, 1883.

White, Andrew Dickson. *Autobiography of Andrew Dickson White*. New York: The Century Co., 1905.

Whitehill, Walter Muir. *Boston, A Topographical History* Cambridge: Harvard University Press, 1968.

Whiting, Emma Mayhew, and Henry Beetle Hough. *Whaling Wives*. Boston: Houghton Mifflin, 1953.

Wilson, Harold F. *Cottages and Commuters, A History of Pitman, New Jersey*. Pitman: Pitman Borough, 1955.

Wright, John F. *Sketches of the Life and Labors of James Quinn*. Cincinnati: Methodist Book Concern, 1851.

INDEX